Lecture Notes in Mathematics 2035

Editors:
J.-M. Morel, Cachan
B. Teissier, Paris

Editors *Mathematical Biosciences Subseries:*
P.K. Maini, Oxford

For further volumes:
http://www.springer.com/series/304

Habib Ammari

Editor

Mathematical Modeling in Biomedical Imaging II

Optical, Ultrasound, and Opto-Acoustic Tomographies

 Springer

Editor
Habib Ammari
École Normale Supérieure
Mathématiques et Applications
45 rue d'Ulm
75005 Paris
France
habib.ammari@ens.fr

ISBN 978-3-642-22989-3   e-ISBN 978-3-642-22990-9
DOI 10.1007/978-3-642-22990-9
Springer Heidelberg Dordrecht London New York

Lecture Notes in Mathematics ISSN print edition: 0075-8434
ISSN electronic edition: 1617-9692

Library of Congress Control Number: 2009932673

Mathematics Subject Classification (2011): 35-XX; 65-XX; 92-XX

Printed on acid-free paper

Springer is part of Springer Science+Business Media (www.springer.com)

Preface

The aim of this volume is to report recent mathematical and computational advances in optical, ultrasound, and photo-acoustic (also called opto-acoustic) tomographies. The volume outlines the state-of-the-art and future directions in optical and ultrasound imaging. It provides some of the most recent mathematical and computational tools in these fields. It is particularly suitable for researchers and graduate students in applied mathematics and in biomedical engineering.

Ultrasound imaging is based on the detection of mechanical properties (acoustic impedance) in biological soft tissues. It can provide good spatial resolution because of its millimetric wavelength and weak scattering at MHz frequencies. However, soft-tissue contrast is relatively poor. Optical tomography is a biomedical imaging modality that uses scattered light as a probe of structural variations in the optical properties of tissue. Optical imaging is very sensitive to optical absorption but can only provide a spatial resolution on the order of 1 cm at cm depths.

Photo-acoustic imaging is a promising new biomedical imaging modality. It combines both optical and ultrasound approaches to provide images of optical contrasts (based on the optical absorption) with ultrasonic resolution.

The objective of the volume is fourfold: (i) to discuss models for light propagation and present fast algorithms for solving the radiative transfer equation; (ii) to provide efficient weighted-migration algorithms for detecting acoustic anomalies and to investigate the coherent interferometric imaging strategy in the acoustic wave propagation regime relevant for biomedical applications; (iii) to explain some experimental setups and survey mathematical inversion techniques in photo-acoustic tomography; (iv) to compensate the effect of acoustic attenuation in purely acoustic as well as photo-acoustic imaging.

The book is organized as follows. Chapter 1 outlines recent mathematical advances in the image reconstruction problem of optical tomography. It gives models for light propagation on microscopic, mesoscopic and macroscopic scales. The mathematical formulation of the corresponding forward problem is dictated primarily by spatial scale, ranging from the Maxwell equations at the microscale, to the radiative transport equation at the mesoscale, and to diffusion theory at the macroscale. The corresponding inverse problem that arises at each of these scales of reconstructing the optical properties of a medium of interest from boundary measurements is considered. An emphasis is put on direct methods for image reconstruction.

Chapter 2 reports recent mathematical and computational advances in the image reconstruction problem of ultrasound tomography. It discusses expansion methods and reverse migration algorithms for detecting acoustic anomalies. When the acoustic medium is randomly heterogeneous, travel times cannot be known with accuracy so that images obtained with reverse migration are noisy and not statistically stable, that is, they change with the realization of the random medium. Coherent interferometry (CINT) has been shown to achieve a good compromise between resolution and deblurring for imaging in noisy environments. CINT consists of backpropagating the cross correlations of the recorded signals over appropriate space-time or space-frequency windows rather than the signals themselves. Chapter 2 provides a CINT strategy in the acoustic wave propagation regime relevant for biomedical applications.

Chapter 3 is devoted to photo-acoustic tomography. Photo-acoustic tomography utilizes opto-acoustic effects of an absorbing medium; when a sample is illuminated by a short electromagnetic pulse, such as visible light, or radio wave, it induces an acoustic wave. The generated pressure field of the acoustic wave depends on the spatially varying absorption density of the sample.

In photo-acoustic imaging the goal is to recover the density function from measurement data of the acoustic pressure taken outside the illuminated sample. Chapter 3 outlines the principles of photo-acoustic tomography. It presents a few reconstruction algorithms which allow to correct the effects of imposed boundary conditions and of acoustic attenuation in photo-acoustic image reconstruction. In photo-acoustic imaging, if the medium is acoustically homogeneous and has the same acoustic properties as the free space, then the boundary of the object plays no role and the optical properties of the medium can be extracted from measurements of the pressure wave by inverting a spherical Radon transform. However, if a boundary condition has to be imposed on the pressure field, then there is no explicit inversion formula. Using a quite simple duality approach, one can still reconstruct the optical absorption coefficient.

Chapter 3 investigates quantitative photo-acoustic imaging in the case of a bounded medium with imposed boundary conditions. It proposes a geometric-control approach to deal with the case of limited view measurements. For small optical absorbers in a non-absorbing background, Chapter 3 provides adapted algorithms to identify the locations of the absorbers and estimate their absorbed energy. An efficient approach in the case of extended optical sources and attenuating acoustic background is also designed. By testing the boundary measurements against an appropriate family of functions, one can access the Radon transform of the initial condition, and thus recovers quantitatively any initial condition for the photo-acoustic problem. Chapter 3 shows how to compensate the effect of acoustic attenuation on image quality for extended absorbers.

Chapter 4 goes further on the investigation of the effect of attenuation on photo-acoustic imaging. The existing attenuation models are reviewed in some detail and their causality is discussed; which is an essential property for algorithms for inversion with attenuated data. Then, it surveys causality properties of common attenuation models. Integro-differential equations which the attenuated waves are satisfying are derived. In addition Chapter 4 shows the ill–conditionness of the inverse problem for calculating the non-attenuated wave from the attenuated one.

Chapter 5 is devoted to quantitative photo-acoustic tomography. The problem reduces to reconstruct optical maps, particularly the absorption coefficient, from the deposited optical energy. To reconstruct different maps, there are three methodologies. First, using single optical illumination and assuming that the scattering map is known, one can recover the absorption map; second, using multi-wavelength illuminations and assuming the spectral model of optical coefficients, both absorption and scattering maps can be obtained; third, using multiple optical illumination both absorption and scattering maps can be recovered. Chapter 5 mainly focuses on the third strategy. It provides an efficient algorithm for large-scale three-dimensional quantitative photo-acoustic reconstructions to simultaneously reconstruct the absorption coefficient and scattering coefficient.

The partial support by the ANR project EchoScan (AN-06-Blan-0089) and the ERC Advanced Grant Project MULTIMOD–267184 are acknowledged. We also thank the staff at the Institute Henri Poincaré.

Paris   *Habib Ammari*

Contents

1

Direct Reconstruction Methods in Optical Tomography

John C. Schotland

Department of Mathematics, University of Michigan, 503 Thompson Street, Ann Arbor, MI 48109-1340, USA `schotland@umich.edu`

Summary. The aim of this chapter is to present the essential physical ideas that are needed to describe the propagation of light in a random medium. We also discuss various direct reconstruction methods for several inverse problems in optical tomography at mesoscopic and macroscopic scales.

1.1 Introduction

Optical tomography is a biomedical imaging modality that uses scattered light as a probe of structural variations in the optical properties of tissue [1]. In a typical experiment, a highly-scattering medium is illuminated by a narrow collimated beam and the light that propagates through the medium is collected by an array of detectors. In first generation systems, the sources and detectors are coupled to the medium by means of optical fibers. The number of measurements that can be obtained in this manner varies from 10^2 to 10^4 source-detector pairs. More recently, noncontact imaging systems have been introduced, wherein a scanned beam and a lens-coupled CCD camera is employed to replace the illumination and detection fiber-optics of first-generation systems, as shown in Fig. 1.1. Using such a noncontact method, extremely large data sets of order 10^8 measurements can be obtained.

The inverse problem of optical tomography is to reconstruct the optical properties of a medium of interest from boundary measurements. The mathematical formulation of the corresponding forward problem is dictated primarily by spatial scale, ranging from the Maxwell equations at the microscale, to the radiative transport equation at the mesoscale, and to diffusion theory at the macroscale. The standard approach to the inverse problem is framed in terms of nonlinear optimization. It is important to note that although optimization methods are extremely flexible, they lead to iterative algorithms with very high computational cost. Direct reconstruction methods offer an alternative approach that can fill this gap. By direct reconstruction, we mean the use of inversion formulas and associated fast algorithms.

H. Ammari (ed.), *Mathematical Modeling in Biomedical Imaging II*,
Lecture Notes in Mathematics 2035, DOI 10.1007/978-3-642-22990-9_1,
© Springer-Verlag Berlin Heidelberg 2012

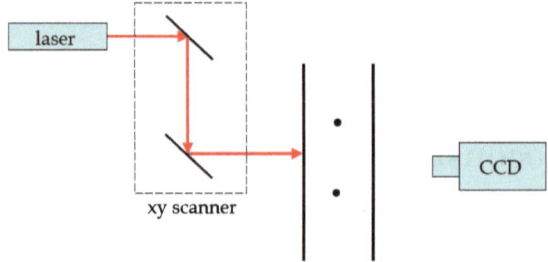

Fig. 1.1. A noncontact optical tomography system

This chapter is organized as follows. In Sect. 1.2 we present the essential physical ideas that are needed to describe the propagation of light in a random medium. We also describe the scattering theory of diffuse waves within radiative transport theory. In Sect. 1.3 we describe various direct reconstruction methods for several inverse problems in optical tomography at mesoscopic and macroscopic scales.

1.2 Forward Problems

The mathematical description of light propagation in random media changes according to the length scale of interest [2]. The Maxwell equations are valid on microscopic scales. The mesoscale, in which the characteristic scale is set by the scattering length, is described by the radiative transport equation (RTE). Finally, we consider the macroscale, which is described by the diffusion approximation to the RTE.

1.2.1 Radiative Transport

In radiative transport theory, the propagation of light through a material medium is formulated in terms of a conservation law that accounts for gains and losses of photons due to scattering and absorption [3,4]. The fundamental quantity of interest is the specific intensity $I(\boldsymbol{r}, \hat{\mathbf{s}})$, defined as the intensity at the position $\boldsymbol{r}$ in the direction $\hat{\mathbf{s}}$. The specific intensity obeys the radiative transport equation (RTE):

$$\hat{\mathbf{s}} \cdot \nabla I + (\mu_a + \mu_s)I = \mu_s \int p(\hat{\mathbf{s}}', \hat{\mathbf{s}}) I(\boldsymbol{r}, \hat{\mathbf{s}}') d\hat{\mathbf{s}}' , \quad \boldsymbol{r} \in \Omega , \qquad (1.1)$$

where μ_a and μ_s are the absorption and scattering coefficients. The specific intensity also satisfies the half-range boundary condition

$$I(\boldsymbol{r}, \hat{\mathbf{s}}) = I_{\text{inc}}(\boldsymbol{r}, \hat{\mathbf{s}}) , \quad \hat{\mathbf{s}} \cdot \hat{\boldsymbol{\nu}} < 0 , \quad \boldsymbol{r} \in \partial\Omega , \qquad (1.2)$$

where $\hat{\nu}$ is the outward unit normal to $\partial\Omega$ and I_{inc} is the incident specific intensity at the boundary. The above choice of boundary condition guarantees the uniqueness of solutions to the RTE [4]. The phase function p is symmetric with respect to interchange of its arguments and obeys the normalization condition

$$\int p(\hat{s}, \hat{s}')d\hat{s}' = 1 \ , \tag{1.3}$$

for all $\hat{s}$. We will often assume that $p(\hat{s}, \hat{s}')$ depends only upon the angle between $\hat{s}$ and $\hat{s}'$, which holds for scattering by spherically-symmetric particles. Note that the choice $p = 1/(4\pi)$ corresponds to isotropic scattering. For scattering that is strongly peaked in the forward direction ($\hat{s} \cdot \hat{s}' \approx 1$), an asymptotic expansion of the right hand side of (1.1) may be performed [5]. This leads to the Fokker-Planck form of the RTE

$$\hat{s} \cdot \nabla I + (\mu_a + \mu_s)I = \mu_s \frac{1}{2}(1 - g)\Delta_{\hat{s}}I \ , \tag{1.4}$$

where $\Delta_{\hat{s}}$ is the Laplacian on the two-dimensional unit sphere S^2 and $g \approx 1$ is the anisotropy of scattering, as given by (1.60) in Sect. (1.2.4).

The total power P passing through a surface Σ is related to the specific intensity by

$$P = \int_{\Sigma} d\mathbf{r} \int d\hat{s} I(\mathbf{r}, \hat{s})\hat{s} \cdot \hat{\nu} \ . \tag{1.5}$$

The energy density Φ is obtained by integrating out the angular dependence of the specific intensity:

$$\Phi(\mathbf{r}) = \frac{1}{c} \int I(\mathbf{r}, \hat{s})d\hat{s} \ . \tag{1.6}$$

We note that the RTE allows for the addition of intensities. As a result, it cannot explain certain wavelike phenomena.

From Waves to Transport

The RTE can be derived by considering the high-frequency asymptotics of wave propagation in a random medium. We briefly recall the main ideas in the context of monochromatic scalar waves. The general theory for vector electromagnetic waves is presented in [6].

We begin by recalling that, within the scalar approximation to the Maxwell equations, the electric field U, in an inhomogeneous medium with a position-dependent permittivity ε, satisfies the time-independent wave equation

$$\nabla^2 U(\mathbf{r}) + k_0^2 \varepsilon(\mathbf{r})U(\mathbf{r}) = 0 \ , \tag{1.7}$$

where k_0 is the free-space wavenumber. The conservation of energy is governed by the relation

$$\nabla \cdot \boldsymbol{J} + \frac{4\pi k_0}{c} \mathrm{Im}\,(\varepsilon)\,I = 0 \ , \tag{1.8}$$

where the energy current density $\boldsymbol{J}$ is defined by

$$\boldsymbol{J} = \frac{1}{2ik_0}\,(U^*\nabla U - U\nabla U^*) \tag{1.9}$$

and the intensity I is given by

$$I = \frac{c}{4\pi}|U|^2 \ . \tag{1.10}$$

Note that $\boldsymbol{J}$ plays the role of the Poynting vector in the scalar theory.

We assume that the random medium is statistically homogeneous and that the susceptibility η is a Gaussian random field such that

$$\langle\eta(\boldsymbol{r})\rangle = 0 \ , \quad \langle\eta(\boldsymbol{r})\eta(\boldsymbol{r}')\rangle = C(|\boldsymbol{r}-\boldsymbol{r}'|) \ , \tag{1.11}$$

where C is the two-point correlation function and $\langle\cdots\rangle$ denotes statistical averaging. Let L denote the propagation distance of the wave. At high frequencies, L is large compared to the wavelength and we introduce a small parameter $\epsilon = 1/(k_0 L) \ll 1$. We suppose that the fluctuations in η are weak so that C is of the order $O(\epsilon)$. We then rescale the spatial variable according to $\boldsymbol{r} \to \boldsymbol{r}/\epsilon$ and define the scaled field $U_\epsilon(\boldsymbol{r}) = U(\boldsymbol{r}/\epsilon)$, so that (1.7) becomes

$$\epsilon^2\nabla^2 U_\epsilon(\boldsymbol{r}) + U_\epsilon(\boldsymbol{r}) = -4\pi\sqrt{\epsilon}\eta\,(\boldsymbol{r}/\epsilon)\,U_\epsilon(\boldsymbol{r}) \ . \tag{1.12}$$

Here we have introduced a rescaling of η to be consistent with the assumption that the fluctuations are of strength $O(\epsilon)$.

Although (1.8) gives some indication of how the intensity of the field is distributed in space, it does not prescribe how the intensity propagates. To overcome this difficulty, we introduce the Wigner distribution $W_\epsilon(\boldsymbol{r}, \boldsymbol{k})$, which is a function of the position $\boldsymbol{r}$ and the wave vector $\boldsymbol{k}$:

$$W_\epsilon(\boldsymbol{r}, \boldsymbol{k}) = \int d\boldsymbol{R}\,e^{i\boldsymbol{k}\cdot\boldsymbol{R}}U_\epsilon\left(\boldsymbol{r} - \frac{1}{2}\epsilon\boldsymbol{R}\right)U_\epsilon^*\left(\boldsymbol{r} + \frac{1}{2}\epsilon\boldsymbol{R}\right) \ . \tag{1.13}$$

The Wigner distribution has several important properties. It is real-valued and is related to the intensity and energy current density by the formulas

$$I = \frac{c}{4\pi}\int \frac{d\boldsymbol{k}}{(2\pi)^3}W_\epsilon(\boldsymbol{r}, \boldsymbol{k}) \ , \quad \boldsymbol{J} = \int \frac{d\boldsymbol{k}}{(2\pi)^3}\boldsymbol{k}W_\epsilon(\boldsymbol{r}, \boldsymbol{k}) \ . \tag{1.14}$$

Making use of (1.12), it can be seen that the Wigner distribution obeys the equation

$$\boldsymbol{k} \cdot \nabla_r W_\epsilon + i\frac{2\pi}{\sqrt{\epsilon}} \int dq e^{-i\boldsymbol{q}\cdot\boldsymbol{x}/\epsilon} \tilde{\eta}(\boldsymbol{q}) \left(W_\epsilon(\boldsymbol{r}, \boldsymbol{k} + \frac{1}{2}\boldsymbol{q}) - W_\epsilon(\boldsymbol{r}, \boldsymbol{k} - \frac{1}{2}\boldsymbol{q}) \right) = 0 \ ,$$
$$(1.15)$$

where we have assumed that η is real-valued and $\tilde{\eta}$ denotes the Fourier transform of η which is defined by

$$\tilde{\eta}(\boldsymbol{q}) = \int d\boldsymbol{r} e^{i\boldsymbol{q}\cdot\boldsymbol{r}} \eta(\boldsymbol{r}) \ . \qquad (1.16)$$

We now consider the asymptotics of the Wigner function in the homogenization limit $\epsilon \to 0$. This corresponds to the regime of high-frequencies and weak fluctuations. We proceed by introducing a two-scale expansion for W_ϵ of the form

$$W_\epsilon(\boldsymbol{r}, \boldsymbol{r}', \boldsymbol{k}) = W_0(\boldsymbol{r}, \boldsymbol{k}) + \sqrt{\epsilon} W_1(\boldsymbol{r}, \boldsymbol{r}', \boldsymbol{k}) + \epsilon W_2(\boldsymbol{r}, \boldsymbol{r}', \boldsymbol{k}) + \cdots \ , \qquad (1.17)$$

where $\boldsymbol{r}' = \boldsymbol{r}/\epsilon$ is a fast variable. By averaging over the fluctuations on the fast scale, it is possible to show that $\langle W_0 \rangle$, which we denote by W, obeys the equation

$$\boldsymbol{k} \cdot \nabla_r W = \int d\boldsymbol{k}' \tilde{C}(\boldsymbol{k} - \boldsymbol{k}') \delta(k^2 - k'^2) \left(W(\boldsymbol{r}, \boldsymbol{k}') - W(\boldsymbol{r}, \boldsymbol{k}) \right) \ . \qquad (1.18)$$

Evidently, (1.18) has the form of a time independent transport equation. The role of the delta function is to conserve momentum, making it possible to view W as a function of position and the direction $\boldsymbol{k}/|\boldsymbol{k}|$. We note that the phase function and scattering coefficient are related to statistical properties of the random medium, as reflected in the appearance of the correlation function C in (1.18). If the medium is composed of spatially uncorrelated point particles with number density ρ, then

$$\mu_a = \rho\sigma_a \ , \quad \mu_s = \rho\sigma_s \ , \quad p = \frac{d\sigma_s}{d\Omega} \bigg/ \sigma_s \ , \qquad (1.19)$$

where σ_a and σ_s are the absorption and scattering cross sections of the particles and $d\sigma_s/d\Omega$ is the differential scattering cross section. Note that σ_a, σ_s and p are wavelength dependent quantities.

1.2.2 Collision Expansion

The RTE (1.1), obeying the boundary condition (1.2), is equivalent to the integral equation

$$I(\boldsymbol{r}, \hat{\boldsymbol{s}}) = I_0(\boldsymbol{r}, \hat{\boldsymbol{s}}) + \int G_0(\boldsymbol{r}, \hat{\boldsymbol{s}}; \boldsymbol{r}', \hat{\boldsymbol{s}}') \mu_s(\boldsymbol{r}') p(\hat{\boldsymbol{s}}', \hat{\boldsymbol{s}}'') I(\boldsymbol{r}', \hat{\boldsymbol{s}}'') d\boldsymbol{r}' d\hat{\boldsymbol{s}}' d\hat{\boldsymbol{s}}'' \ . \qquad (1.20)$$

Here I_0 is the unscattered (ballistic) specific intensity, which satisfies the equation

$$[\hat{\mathbf{s}} \cdot \nabla + \mu_a + \mu_s] I_0 = 0 , \tag{1.21}$$

and G_0 is the ballistic Green's function

$$G_0(\mathbf{r}, \hat{\mathbf{s}}; \mathbf{r}', \hat{\mathbf{s}}') = g(\mathbf{r}, \mathbf{r}')\delta \left(\hat{\mathbf{s}}' - \frac{\mathbf{r} - \mathbf{r}'}{|\mathbf{r} - \mathbf{r}'|} \right) \delta(\hat{\mathbf{s}} - \hat{\mathbf{s}}') , \tag{1.22}$$

where

$$g(\mathbf{r}, \mathbf{r}') = \frac{1}{|\mathbf{r} - \mathbf{r}'|^2} \exp \left[-\int_0^{|\mathbf{r}-\mathbf{r}'|} \mu_t \left(\mathbf{r}' + \ell \frac{\mathbf{r} - \mathbf{r}'}{|\mathbf{r} - \mathbf{r}'|} \right) d\ell \right] , \tag{1.23}$$

and the extinction coefficient $\mu_t = \mu_a + \mu_s$. Note that if a narrow collimated beam of intensity I_{inc} is incident on the medium at the point $\mathbf{r}_0$ in the direction $\hat{\mathbf{s}}_0$, then $I_0(\mathbf{r}, \hat{\mathbf{s}})$ is given by

$$I_0(\mathbf{r}, \hat{\mathbf{s}}) = I_{\text{inc}} G_0(\mathbf{r}, \hat{\mathbf{s}}; \mathbf{r}_0, \hat{\mathbf{s}}_0) , \tag{1.24}$$

To derive the collision expansion, we iterate (1.20) starting from $I^{(0)} = I_0$ and obtain

$$I(\mathbf{r}, \hat{\mathbf{s}}) = I^{(0)}(\mathbf{r}, \hat{\mathbf{s}}) + I^{(1)}(\mathbf{r}, \hat{\mathbf{s}}) + I^{(2)}(\mathbf{r}, \hat{\mathbf{s}}) + \cdots , \tag{1.25}$$

where each term of the series is given by

$$I^{(n)}(\mathbf{r}, \hat{\mathbf{s}}) = \int d\mathbf{r}' d\hat{\mathbf{s}}' d\hat{\mathbf{s}}'' G_0(\mathbf{r}, \hat{\mathbf{s}}; \mathbf{r}', \hat{\mathbf{s}}')\mu_s(\mathbf{r}')p(\hat{\mathbf{s}}', \hat{\mathbf{s}}'')I^{(n-1)}(\mathbf{r}', \hat{\mathbf{s}}'') , \tag{1.26}$$

with $n = 1, 2, \ldots$. The above series is the analog of the Born series for the RTE, since each term accounts for successively higher orders of scattering.

It is instructive to examine the expression for $I^{(1)}$, which is the contribution to the specific intensity from single scattering:

$$I^{(1)}(\mathbf{r}, \hat{\mathbf{s}}) = \int d\mathbf{r}' d\hat{\mathbf{s}}' d\hat{\mathbf{s}}'' G_0(\mathbf{r}, \hat{\mathbf{s}}; \mathbf{r}', \hat{\mathbf{s}}')\mu_s(\mathbf{r}')p(\hat{\mathbf{s}}', \hat{\mathbf{s}}'')I_0(\mathbf{r}', \hat{\mathbf{s}}'') . \tag{1.27}$$

It follows from (1.24) that the change in intensity $\delta I = I - I^{(0)}$ measured by a point detector at $\mathbf{r}_2$ in the direction $\hat{\mathbf{s}}_2$, due to a unit-amplitude point source at $\mathbf{r}_1$ in the direction $\hat{\mathbf{s}}_1$ is given by

$$\delta I(\mathbf{r}_1, \hat{\mathbf{s}}_1; \mathbf{r}_2, \hat{\mathbf{s}}_2) = p(\hat{\mathbf{s}}_1, \hat{\mathbf{s}}_2) \int_0^\infty dR R^2 g(\mathbf{r}_2, \mathbf{r}_1 + R\hat{\mathbf{s}}_1)g(\mathbf{r}_1 + R\hat{\mathbf{s}}_1, \mathbf{r}_1)$$
$$\times \delta \left(\frac{\mathbf{r}_2 - \mathbf{r}_1 - R\hat{\mathbf{s}}_1}{|\mathbf{r}_2 - \mathbf{r}_1 - R\hat{\mathbf{s}}_1|} - \hat{\mathbf{s}}_2 \right) \mu_s(\mathbf{r}_1 + R\hat{\mathbf{s}}_1) . \tag{1.28}$$

Suppose that r_1 and r_2 are located on the boundary of a bounded domain and $\hat{s}_1$ points into and $\hat{s}_2$ points out of the domain. Then the rays in the directions $\hat{s}_1$ and $\hat{s}_2$ must intersect at a point $\boldsymbol{R}$ that lies in the interior of the domain. In addition, the delta function in (1.28) implements the constraint that $\hat{s}_1$, $\hat{s}_2$ and $r_1 - r_2$ all lie in the same plane as shown in Fig. 1.2. Using this fact and carrying out the integration in (1.28), we find that

$$
\delta I(r_1, \hat{s}_1; r_2, \hat{s}_2) \propto \exp\left[- \int_0^{L_1} \mu_t(r_1 + \ell\hat{s}_1) d\ell \right.
$$
$$
\left. - \int_0^{L_2} \mu_t(\boldsymbol{R} + \ell\hat{s}_2) d\ell \right], \qquad (1.29)
$$

where $L_1 = |\boldsymbol{R} - r_1|$, $L_2 = |\boldsymbol{R} - r_2|$ and we have omitted overall geometric prefactors. Note that the argument of the exponential corresponds to the integral of μ_t along the broken ray which begins at r_1, passes through $\boldsymbol{R}$, and terminates at r_2. The significance of such broken rays will be discussed in Sect. 1.3.3.

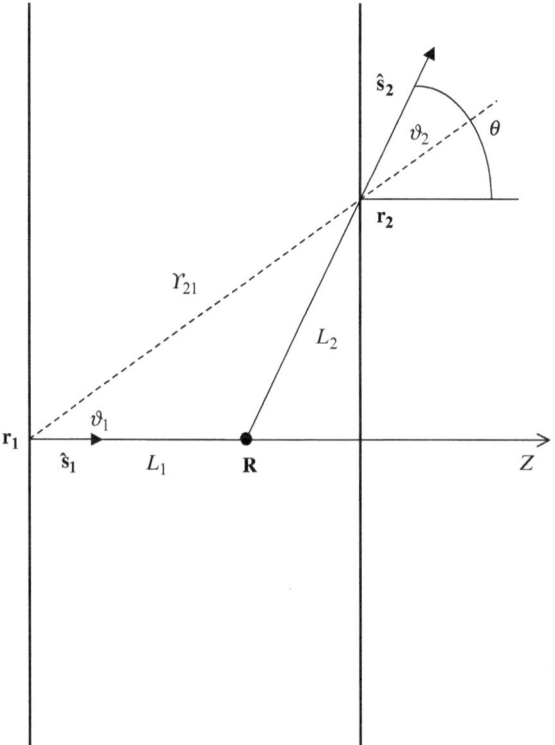

Fig. 1.2. Illustrating the geometry of (1.28)

The terms in the collision expansion can be classified by their smoothness. The lowest order term is the most singular. In the absence of scattering, according to (1.23), this term leads to a Radon transform relationship between the absorption coefficient and the specific intensity, under that condition that the source and detector are collinear. Inversion of the Radon transform is the basis for optical projection tomography [7,8]. The first order term is also singular, as is evident from the presence of a delta function in (1.28). Terms of higher order are of increasing smoothness. This observation has been exploited to prove uniqueness of the inverse transport problem and to study its stability. A comprehensive review is presented in [9].

The above discussion has implicitly assumed that the angular dependence of the specific intensity is measurable. In practice, such measurements are extremely difficult to obtain. The experimentally measurable intensity is often an angular average of the specific intensity over the aperture of an optical system. The effect of averaging is to remove the singularities that are present in the specific intensity. The resulting inverse problem is then highly ill-posed [10]. To illustrate this point, we observe that it follows from (1.6) and (1.20) that the energy density Φ obeys the integral equation

$$\Phi(\boldsymbol{r}) = \Phi^{(0)}(\boldsymbol{r}) + \frac{1}{4\pi} \int d\boldsymbol{r}' \frac{e^{-\mu_t|\boldsymbol{r}-\boldsymbol{r}'|}}{|\boldsymbol{r}-\boldsymbol{r}'|^2} p(\boldsymbol{r}')\Phi(\boldsymbol{r}') \ , \tag{1.30}$$

where the scattering is assumed to be isotropic and

$$\Phi^{(0)}(\boldsymbol{r}) = \frac{1}{c} \int I^{(0)}(\boldsymbol{r}, \hat{\mathbf{s}}) d\hat{\mathbf{s}} \ . \tag{1.31}$$

Here the kernel appearing in (1.30) is smoothing with a Fourier transform that decays algebraically at high frequencies.

1.2.3 Transport Regime

We now develop the scattering theory for the RTE in an inhomogeneously absorbing medium. We refer to Chap. 5 for numerical solutions to the RTE and its diffusion approximation.

We proceed by decomposing μ_a into a constant part $\bar{\mu}_a$ and a spatially varying part $\delta\mu_a$:

$$\mu_a(\boldsymbol{r}) = \bar{\mu}_a + \delta\mu_a(\boldsymbol{r}) \ . \tag{1.32}$$

The stationary form of the RTE (1.1) can be rewritten in the form

$$\hat{\mathbf{s}} \cdot \nabla I + \bar{\mu}_t I - \mu_s \int p(\hat{\mathbf{s}}', \hat{\mathbf{s}}) I(\boldsymbol{r}, \hat{\mathbf{s}}') d\hat{\mathbf{s}}' = -\delta\mu_a(\boldsymbol{r})I \ , \tag{1.33}$$

where $\bar{\mu}_t = \bar{\mu}_a + \mu_s$. The solution to (1.33) is given by

$$I(\boldsymbol{r}, \hat{\mathbf{s}}) = I_i(\boldsymbol{r}, \hat{\mathbf{s}}) - \int d\boldsymbol{r}' d\hat{\mathbf{s}}' G(\boldsymbol{r}, \hat{\mathbf{s}}; \boldsymbol{r}', \hat{\mathbf{s}}') \delta\mu_a(\boldsymbol{r}')I(\boldsymbol{r}', \hat{\mathbf{s}}') \ , \tag{1.34}$$

where G denotes the Green's function for a homogeneous medium with absorption $\bar{\mu}_a$ and I_i is the incident specific intensity. Equation (1.34) is the analog of the Lippmann-Schwinger equation for the RTE. It describes the "multiple scattering" of the incident specific intensity from inhomogeneities in $\delta\mu_a$. If only one absorption event is considered, then the intensity I on the right hand side of (1.34) can be replaced by the incident intensity I_i. This result describes the linearization of the integral (1.34) with respect to $\delta\mu_a$. If the incident field is generated by a point source at r_1 pointing in the direction $\hat{s}_1$, then the change in specific intensity due to spatial fluctuations in absorption δI can be obtained from the relation

$$\delta I(r_1, \hat{s}_1; r_2, \hat{s}_2) = I_0 \int dr d\hat{s} G(r_1, \hat{s}_1; r, \hat{s}) G(r, \hat{s}; r_2, \hat{s}_2) \delta\mu_a(r) . \quad (1.35)$$

Here I_0 denotes the intensity of the source and $r_2, \hat{s}_2$ are the position and orientation of a point detector.

To make further progress requires knowledge of the Green's function for the RTE. The Green's function for an infinite medium with isotropic scattering can be obtained explicitly [4]. In three dimensions we have

$$G(r, \hat{s}; r', \hat{s}') = G_0(r, \hat{s}; r', \hat{s}') + \frac{\mu_s}{4\pi} \int \frac{dk}{(2\pi)^3} e^{ik \cdot (r - r')} \frac{1}{(\mu_t + i\hat{s} \cdot k)(\mu_t + i\hat{s}' \cdot k)}$$

$$\times \frac{1}{1 - \frac{\mu_s}{|k|} \tan^{-1}\left(\frac{|k|}{\mu_t}\right)} , \quad (1.36)$$

where G_0 is the ballistic Green's function which is defined by (1.22). More generally, numerical procedures such as the discrete ordinate method or the P_N approximation may be employed to compute the Green's function in a bounded domain with anisotropic scattering.

The method of rotated reference frames is a spectral method for the computing the Green's function for the three-dimensional RTE in a homogeneous medium with anisotropic scattering and planar boundaries [11]. It is derived by considering the plane-wave modes for the RTE which are of the form

$$I(r, \hat{s}) = A_k(\hat{s}) e^{k \cdot r} , \quad (1.37)$$

where the amplitude A is to be determined. Evidently, the components of the wave vector k cannot be purely real; otherwise the above modes would have exponential growth in the $\hat{k}$ direction. We thus consider evanescent modes with

$$k = iq \pm \sqrt{q^2 + 1/\lambda^2} \, \hat{z} , \quad (1.38)$$

where $q \cdot \hat{z} = 0$ and $k \cdot k \equiv 1/\lambda^2$. These modes are oscillatory in the transverse direction and decay in the $\pm z$-directions. By inserting (1.37) into the RTE (1.1), we find that $A_k(\hat{s})$ satisfies the equation

$$(\hat{s} \cdot k + \mu_a + \mu_s) A_k(\hat{s}) = \mu_s \int p(\hat{s}, \hat{s}') A_k(\hat{s}') d\hat{s}' . \quad (1.39)$$

To solve the eigenproblem defined by (1.39) it will prove useful to expand $A_{\boldsymbol{k}}(\hat{\mathbf{s}})$ into a basis of spherical functions defined in a rotated reference frame whose z-axis coincides with the direction $\hat{\boldsymbol{k}}$. We denote such functions by $Y_{lm}(\hat{\mathbf{s}}; \hat{\boldsymbol{k}})$ and define them via the relation

$$Y_{lm}(\hat{\mathbf{s}}; \hat{\boldsymbol{k}}) = \sum_{m'=-l}^{l} D^{l}_{mm'}(\varphi, \theta, 0) Y_{lm'}(\hat{\mathbf{s}}) , \qquad (1.40)$$

where $Y_{lm}(\hat{\mathbf{s}})$ are the spherical harmonics defined in the laboratory frame, $D^{l}_{mm'}$ is the Wigner D-function and φ, θ are the polar angles of $\hat{\boldsymbol{k}}$ in the laboratory frame. We thus expand $A_{\boldsymbol{k}}$ as

$$A_{\boldsymbol{k}}(\hat{\mathbf{s}}) = \sum_{l,m} C_{lm} Y_{lm}(\hat{\mathbf{s}}; \hat{\boldsymbol{k}}) , \qquad (1.41)$$

where the coefficients C_{lm} are to be determined. Note that since the phase function $p(\hat{\mathbf{s}}, \hat{\mathbf{s}}')$ is invariant under simultaneous rotation of $\hat{\mathbf{s}}$ and $\hat{\mathbf{s}}'$, it may be expanded into rotated spherical functions according to

$$p(\hat{\mathbf{s}}, \hat{\mathbf{s}}') = \sum_{l,m} p_{l} Y_{lm}(\hat{\mathbf{s}}; \hat{\boldsymbol{k}}) Y^{*}_{lm}(\hat{\mathbf{s}}'; \hat{\boldsymbol{k}}) , \qquad (1.42)$$

where the expansion coefficients p_{l} are independent of $\hat{\boldsymbol{k}}$. An alternative approach that may be used to solve the eigenproblem (1.39) is to employ the method of discrete ordinates [12–14].

Substituting (1.41) into (1.39) and making use of the orthogonality properties of the spherical functions, we find that the coefficients C_{lm} satisfy the equation

$$\sum_{l',m'} R^{lm}_{l'm'} C_{l'm'} = \lambda \sigma_{l} C_{lm} . \qquad (1.43)$$

Here the matrix R is defined by

$$R^{lm}_{l'm'} = \int d\hat{\mathbf{s}} \, \hat{\mathbf{s}} \cdot \hat{\boldsymbol{k}} Y_{lm}(\hat{\mathbf{s}}; \hat{\boldsymbol{k}}) Y^{*}_{l'm'}(\hat{\mathbf{s}}; \hat{\boldsymbol{k}}) \qquad (1.44)$$

$$= \delta_{mm'} \left(b_{lm} \delta_{l',l-1} + b_{l+1,m} \delta_{l',l-1} \right) ,$$

where

$$b_{lm} = \sqrt{(l^2 - m^2)/(4l^2 - 1)} , \qquad (1.45)$$

and

$$\sigma_{l} = \mu_{a} + \mu_{s}(1 - p_{l}) . \qquad (1.46)$$

Equation (1.43) defines a generalized eigenproblem which can be transformed into a standard eigenproblem as follows. Define the diagonal matrix $S^{lm}_{l'm'} = \delta_{mm'}\delta_{ll'}\sqrt{\sigma_{l}}$. Note that $\sigma_{l} > 0$ since $p_{l} \leq 1$ and thus S is well defined. We then pre and post multiply R by S^{-1} and find that $W\psi = \lambda\psi$ where

$W = S^{-1}RS^{-1}$ and $\psi = SC$. It can be shown that W is symmetric and block tridiagonal with both a discrete and continuous spectrum of eigenvalues λ_μ and a corresponding complete orthonormal set of eigenvectors ψ_μ, indexed by μ [11]. We thus see that the modes (1.37), which are labeled by μ, the transverse wave vector $\boldsymbol{q}$, and the direction of decay, are of the form

$$I_{\boldsymbol{q}\mu}^{\pm}(\boldsymbol{r},\hat{\mathbf{s}}) = \sum_{l,m}\sum_{m'}\frac{1}{\sqrt{\sigma_l}}\psi_{lm}^{\mu}D_{mm'}^{l}(\varphi,\theta,0)Y_{lm'}(\hat{\mathbf{s}})e^{i\boldsymbol{q}\cdot\boldsymbol{\rho}\mp Q_\mu(\boldsymbol{q})z} , \qquad (1.47)$$

where

$$Q_\mu(\boldsymbol{q}) = \sqrt{q^2 + 1/\lambda_\mu^2} . \qquad (1.48)$$

The Green's function for the RTE in the $z \geq 0$ half-space may be constructed as a superposition of the above modes:

$$G(\boldsymbol{r},\hat{\mathbf{s}};\boldsymbol{r}',\hat{\mathbf{s}}') = \int \frac{d\boldsymbol{q}}{(2\pi)^2}\sum_\mu \mathcal{A}_{\boldsymbol{q}\mu}I_{\boldsymbol{q}\mu}^{\pm}(\boldsymbol{r},\hat{\mathbf{s}})I_{-\boldsymbol{q}\mu}^{\mp}(\boldsymbol{r}',-\hat{\mathbf{s}}) , \qquad (1.49)$$

where the upper sign is chosen if $z > z'$, the lower sign is chosen if $z < z'$ and the coefficients $\mathcal{A}_{\boldsymbol{q}\mu}$ are found from the boundary conditions. Using this result, we see that G can be written as the plane-wave decomposition

$$G(\boldsymbol{r},\hat{\mathbf{s}};\boldsymbol{r}',\hat{\mathbf{s}}') = \int \frac{d\boldsymbol{q}}{(2\pi)^2}\sum_{lm,l'm'}g_{l'm'}^{lm}(z,z';\boldsymbol{q})e^{i\boldsymbol{q}\cdot(\boldsymbol{\rho}-\boldsymbol{\rho}')}Y_{lm}(\hat{\mathbf{s}})Y_{l'm'}^{*}(\hat{\mathbf{s}}') ,$$

$$(1.50)$$

where

$$g_{l'm'}^{lm}(z,z';\boldsymbol{q}) = \frac{1}{\sqrt{\sigma_l\sigma_l'}}\sum_\mu\sum_{M,M'}\mathcal{A}_{\boldsymbol{q}\mu}\psi_{lm}^{\mu}\psi_{l'm'}^{\mu} \qquad (1.51)$$

$$\times D_{mM}^{l}(\varphi,\theta,0)D_{m'M'}^{l'}(\varphi,\theta,0)e^{-Q_\mu(\boldsymbol{q})|z-z'|}$$

$$\equiv \sum_\mu \mathcal{B}_{l'm'}^{lm}(\boldsymbol{q},\mu)e^{-Q_\mu(\boldsymbol{q})|z-z'|} , \qquad (1.52)$$

which defines $\mathcal{B}_{l'm'}^{lm}$. It is important to note that the dependence of G on the coordinates $\boldsymbol{r},\boldsymbol{r}'$ and directions $\hat{\mathbf{s}},\hat{\mathbf{s}}'$ is *explicit* and that the expansion is computable for any rotationally invariant phase function.

1.2.4 Diffuse Light

The diffusion approximation (DA) to the RTE is widely used in applications. It is valid in the regime where the scattering length $l_s = 1/\mu_s$ is small compared to the distance of propagation. The standard approach to the DA is through the P_N approximation, in which the angular dependence of the specific intensity is expanded in spherical harmonics. The DA is obtained if the expansion is truncated at first order. The DA may also be derived using asymptotic methods [15]. The advantage of this approach is that it leads to error estimates and treats the problem of boundary conditions for the resulting diffusion equation in a natural way.

The DA holds when the scattering coefficient is large, the absorption coefficient is small, the point of observation is far from the boundary of the medium and the time-scale is sufficiently long. Accordingly, we perform the rescaling

$$\mu_a \to \epsilon\mu_a , \quad \mu_s \to \frac{1}{\epsilon}\mu_s , \tag{1.53}$$

where $\epsilon \ll 1$. Thus the RTE (1.1) becomes

$$\epsilon\hat{s} \cdot \nabla I + \epsilon^2\mu_a I + \mu_s I = \mu_s \int p(\hat{s}, \hat{s}')I(r, \hat{s}')d\hat{s}' . \tag{1.54}$$

We then introduce the asymptotic expansion for the specific intensity

$$I(r, \hat{s}) = I_0(r, \hat{s}) + \epsilon I_1(r, \hat{s}) + \epsilon^2 I_2(r, \hat{s}) + \cdots \tag{1.55}$$

which we substitute into (1.54). Upon collecting terms of $O(1)$, $O(\epsilon)$ and $O(\epsilon^2)$ we have

$$\int p(\hat{s}, \hat{s}')I_0(r, \hat{s}')d\hat{s}' = I_0(r, \hat{s}) , \tag{1.56}$$

$$\hat{s} \cdot \nabla I_0 + \mu_s I_1 = \mu_s \int p(\hat{s}, \hat{s}')I_1(r, \hat{s}')d\hat{s}' , \tag{1.57}$$

$$\hat{s} \cdot \nabla I_1 + \mu_a I_0 + \mu_s I_2 = \mu_s \int p(\hat{s}, \hat{s}')I_2(r, \hat{s}')d\hat{s}' . \tag{1.58}$$

The normalization condition (1.3) forces I_0 to depend only upon the spatial coordinate r. If the phase function $p(\hat{s}, \hat{s}')$ depends only upon the quantity $\hat{s} \cdot \hat{s}'$, it can be seen that

$$I_1(r, \hat{s}) = -\frac{1}{1-g}\hat{s} \cdot \nabla I_0(r) , \tag{1.59}$$

where the anisotropy g is given by

$$g = \int \hat{s} \cdot \hat{s}'p(\hat{s} \cdot \hat{s}')d\hat{s}' , \tag{1.60}$$

with $-1 < g < 1$. Note that $g = 0$ corresponds to isotropic scattering and $g = 1$ to extreme forward scattering. If we insert the above expression for I_1 into (1.58) and integrate over $\hat{s}$, we obtain the diffusion equation for the energy density Φ:

$$-\nabla \cdot [D(r)\nabla\Phi(r, t)] + c\mu_a(r)u(r, t) = 0 , \tag{1.61}$$

where $I_0 = c\Phi/(4\pi)$. Here the diffusion coefficient is defined by

$$D = \frac{1}{3}c\ell^* \,, \quad \ell^* = \frac{1}{(1-g)\mu_t} \,, \tag{1.62}$$

where ℓ^* is known as the transport mean free path. The usual expression for ℓ^* obtained from the P_N method

$$\ell^* = \frac{1}{(1-g)\mu_s + \mu_a} \,, \tag{1.63}$$

is asymptotically equivalent to (1.62) since $\mu_a = \epsilon^2 \mu_s$. The above derivation of the DA holds in an infinite medium. In a bounded domain, it is necessary to account for boundary layers, since the boundary conditions for the diffusion equation and the RTE are not compatible [15].

We now consider the scattering theory of time-harmonic diffuse waves. Assuming an $e^{-i\omega t}$ time-dependence with modulation frequency ω, the energy density obeys the equation

$$-\nabla \cdot [D(\boldsymbol{r})\nabla\Phi(\boldsymbol{r})] + (c\mu_a(\boldsymbol{r}) - i\omega)\Phi(\boldsymbol{r}) = 0 \quad \text{in} \quad \Omega \,, \tag{1.64}$$

In addition to (1.64), the energy density must satisfy the boundary condition

$$\Phi + \ell_{\text{ext}}\hat{\boldsymbol{\nu}} \cdot \nabla\Phi = g \quad \text{on} \quad \partial\Omega \,, \tag{1.65}$$

where g is the source, $\hat{\boldsymbol{\nu}}$ is the outward unit normal to $\partial\Omega$ and the extrapolation length ℓ_{ext} can be computed from radiative transport theory [4]. We note that $\ell_{\text{ext}} = 0$ corresponds to an absorbing boundary and $\ell_{\text{ext}} \to \infty$ to a reflecting boundary. The solution to (1.64) obeys the Lippmann-Schwinger equation

$$\Phi = \Phi_i - GV\Phi \tag{1.66}$$

where Φ_i is the energy density of the incident diffuse wave and G is the Green's function for a homogeneous medium with absorption $\bar{\mu}_a$ and diffusion coefficient D_0. We have introduced the potential operator which is given by

$$V = c\delta\mu_a - \nabla \cdot (\delta D\nabla) \,, \tag{1.67}$$

where $\delta\mu_a = \mu_a - \bar{\mu}_a$ and $\delta D = D - D_0$. The unperturbed Green's function $G(\boldsymbol{r}, \boldsymbol{r}')$ satisfies

$$\left(\nabla^2 - k^2\right) G(\boldsymbol{r}, \boldsymbol{r}') = -\frac{1}{D_0}\delta(\boldsymbol{r} - \boldsymbol{r}') \,, \tag{1.68}$$

where the diffuse wave number k is given by

$$k^2 = \frac{c\bar{\mu}_a - i\omega}{D_0} \,. \tag{1.69}$$

We note here that the fundamental solution to the diffusion equation is given by

$$G(\boldsymbol{r}, \boldsymbol{r}') = \frac{1}{4\pi D} \frac{e^{-k|\boldsymbol{r}-\boldsymbol{r}'|}}{|\boldsymbol{r} - \boldsymbol{r}'|} \ . \tag{1.70}$$

By iterating (1.66) beginning with $\Phi = \Phi_i$, we obtain

$$\Phi = \Phi_i - GV\Phi_i + GVGV\Phi_i + \cdots \ , \tag{1.71}$$

which is the analog of the Born series for diffuse waves.

1.3 Direct Inversion Methods

By direct inversion we mean the use of inversion formulas and associated fast image reconstruction algorithms. In optical tomography, such formulas exist for particular experimental geometries, including those with planar, cylindrical and spherical boundaries. In this section, we consider several different inverse problems organized by length scale. We begin with the macroscopic case and proceed downward.

1.3.1 Diffuse Optical Tomography

Our aim is to reconstruct the absorption and diffusion coefficients of a macroscopic medium from boundary measurements. We first consider the linearized inverse problem in one dimension and then discuss direct methods for linear and nonlinear inversion in higher dimensions.

One-Dimensional Problem

We start by studying the time-dependent inverse problem in one dimension, which illustrates many features of the three-dimensional case. Let Ω be the half-line $x \geq 0$. The energy density Φ obeys the diffusion equation

$$\frac{\partial}{\partial t}\Phi(x,t) = D\frac{\partial^2}{\partial x^2}\Phi(x,t) - c\mu_a(x)\Phi(x,t) \ , \quad x \in \Omega \ , \tag{1.72}$$

where the diffusion coefficient D is assumed to be constant, an assumption that will be relaxed later. The energy density is taken to obey the initial and boundary conditions

$$\Phi(x,0) = \delta(x - x_1) \ , \tag{1.73}$$

$$\Phi(0,t) - \ell_{\text{ext}}\frac{\partial\Phi}{\partial x}(0,t) = 0 \ . \tag{1.74}$$

Here the initial condition imposes the presence of a point source of unit-strength at x_1. Since Φ decays exponentially, we consider for $k \geq 0$ the Laplace transform

$$\Phi(x, k) = \int_0^\infty e^{-k^2 Dt} \Phi(x, t) dt , \qquad (1.75)$$

which obeys the equation

$$-\frac{d^2 \Phi(x)}{dx^2} + k^2 (1 + \eta(x)) \Phi(x) = \frac{1}{D} \delta(x - x_1) , \qquad (1.76)$$

where η is the spatially-varying part of the absorption, which is defined by $\eta = c\mu_a/(Dk^2) - 1$. The solution to the forward problem is given by the integral equation

$$\Phi(x) = \Phi_i(x) - k^2 \int_\Omega G(x, y) \Phi(y) \eta(y) dy , \qquad (1.77)$$

where the Green's function is of the form

$$G(x, y) = \frac{1}{2Dk} \left(e^{-k|x-y|} + \frac{1 - k\ell_{ext}}{1 + k\ell_{ext}} e^{-k|x+y|} \right) , \qquad (1.78)$$

and Φ_i is the incident field, which obeys (1.76) with $\eta = 0$. The above integral equation may be linearized with respect to $\eta(x)$ by replacing u on the right-hand side by u_i. This approximation is accurate when $\mathrm{supp}(\eta)$ and η are small. If we introduce the scattering data $\Phi_s = \Phi_i - \Phi$ and perform the above linearization, we obtain

$$\Phi_s(x_1, x_2) = k^2 \int_\Omega G(x_1, y) G(y, x_2) \eta(y) dy . \qquad (1.79)$$

Here $\Phi_s(x_1, x_2)$ is proportional to the change in intensity due to a point source at x_1 that is measured by a detector at x_2.

In the backscattering geometry, the source and detector are placed at the origin ($x_1 = x_2 = 0$) and (1.79) becomes, upon using (1.78) and omitting overall constants

$$\Phi_s(k) = \int_0^\infty e^{-kx} \eta(x) dx , \qquad (1.80)$$

where the dependence of Φ_s on k has been made explicit. Thus, the linearized inverse problem can be seen to correspond to inverting the Laplace transform of η. Inversion of the Laplace transform is the paradigmatic exponentially ill-posed problem. It can be analyzed following [16]. Equation (1.80) defines an operator $A : \eta \mapsto \Phi_s$ which is bounded and self-adjoint on $L^2([0, \infty])$. The singular functions f and g of A satisfy

$$A^* A f = \sigma^2 f , \quad AA^* g = \sigma^2 g , \qquad (1.81)$$

where σ is the corresponding singular value. In addition, f and g are related by

$$Af = \sigma g , \quad A^* g = \sigma f . \qquad (1.82)$$

If we observe that $A^*A(x, y) = 1/(x+y)$ and use the identity

$$\int_0^\infty \frac{y^a}{1+y} dy = \frac{\pi}{\sin(a+1)\pi} , \quad -1 \le \mathrm{Re}(a) < 0 , \tag{1.83}$$

we see that

$$f_s(x) = g_s^*(x) = \frac{1}{\sqrt{2\pi}} x^{-\frac{1}{2}+is} , \quad s \in \mathbb{R} \tag{1.84}$$

and

$$\sigma_s^2 = \frac{\pi}{\cosh(\pi s)} \sim e^{-\pi|s|} . \tag{1.85}$$

Note that the singular values of A are exponentially small, which gives rise to severe ill-posedness. Using the above, we can write an inversion formula for (1.78) in the form

$$\eta(x) = \int_0^\infty dk \int_{-\infty}^\infty ds R\left(\frac{1}{\sigma_s}\right) f_s(x) g_s^*(k) \Phi_s(k) , \tag{1.86}$$

where the regularizer R has been introduced to control the contribution of small singular values.

Linearized Inverse Problem

We now consider the linearized inverse problem in three dimensions. To indicate the parallels with the one-dimensional case, we will find it convenient to work in the half-space geometry. Extensions to other geometries, including those with planar, cylindrical and spherical boundaries is also possible [17–19]. In particular, we note that the slab geometry is often employed in optical mammography and small animal imaging. As before, we define the scattering data $\Phi_s = \Phi_i - \Phi$. Linearizing (1.66) with respect to $\delta\mu_a$ and δD we find that, up to an overall constant, Φ_s obeys the integral equation

$$\Phi_s(r_1, r_2) = \int dr \, [G(r_1, r)G(r, r_2)c\delta\mu_a(r)$$
$$+ \nabla_r G(r_1, r) \cdot \nabla_r G(r, r_2)\delta D(r)] , \tag{1.87}$$

where r_1 is the position of the source, r_2 is the position of the detector and we have integrated by parts to evaluate the action of the operator V. The Green's function in the half-space $z \ge 0$ is given by the plane-wave decomposition

$$G(r, r') = \int \frac{dq}{(2\pi)^2} g(q; z, z') \exp[iq \cdot (\rho - \rho')] , \tag{1.88}$$

where we have used the notation $r = (\rho, z)$. If either r or r' lies in the plane $z = 0$, then

$$g(q; z, z') = \frac{\ell_{\mathrm{ext}}}{D_0} \frac{\exp[-Q(q)|z - z'|]}{Q(q)\ell_{\mathrm{ext}} + 1} , \tag{1.89}$$

where

$$Q(\boldsymbol{q}) = \sqrt{q^2 + k^2} \; . \tag{1.90}$$

The inverse problem is to recover $\delta\mu_a$ and δD from boundary measurements. To proceed, we introduce the Fourier transform of Φ_s with respect to the source and detector coordinates according to

$$\tilde{\Phi}_s(\boldsymbol{q}_1, \boldsymbol{q}_2) = \int d\boldsymbol{\rho}_1 d\boldsymbol{\rho}_2 e^{i(\boldsymbol{q}_1 \cdot \boldsymbol{\rho}_1 + \boldsymbol{q}_2 \cdot \boldsymbol{\rho}_2)} \Phi_s(\boldsymbol{\rho}_1, 0; \boldsymbol{\rho}_2, 0) \; . \tag{1.91}$$

If we define

$$\psi(\boldsymbol{q}_1, \boldsymbol{q}_2) = (Q(\boldsymbol{q}_1)\ell_{\text{ext}} + 1)(Q(\boldsymbol{q}_2)\ell_{\text{ext}} + 1)\tilde{\Phi}_s(\boldsymbol{q}_1, \boldsymbol{q}_2) \tag{1.92}$$

and make use of (1.88) and (1.87), we find that

$$\psi(\boldsymbol{q}_1, \boldsymbol{q}_2) = \int d\boldsymbol{r} e^{i(\boldsymbol{q}_1 + \boldsymbol{q}_2) \cdot \boldsymbol{\rho}} \left[\kappa_A(\boldsymbol{q}_1, \boldsymbol{q}_2; z)\delta\mu_a(\boldsymbol{r}) + \kappa_D(\boldsymbol{q}_1, \boldsymbol{q}_2; z)\delta D(\boldsymbol{r}) \right] \; . \tag{1.93}$$

Here

$$\kappa_A(\boldsymbol{q}_1, \boldsymbol{q}_2; z) = c \exp[-(Q(\boldsymbol{q}_1) + Q(\boldsymbol{q}_2)) z] \; , \tag{1.94}$$

$$\kappa_D(\boldsymbol{q}_1, \boldsymbol{q}_2; z) = -(\boldsymbol{q}_1 \cdot \boldsymbol{q}_2 + Q(\boldsymbol{q}_1)Q(\boldsymbol{q}_2)) \exp[-(Q(\boldsymbol{q}_1)$$
$$+ Q(\boldsymbol{q}_2))z] \; . \tag{1.95}$$

We now change variables according to

$$\boldsymbol{q}_1 = \boldsymbol{q} + \boldsymbol{p}/2 \; , \quad \boldsymbol{q}_2 = \boldsymbol{q} - \boldsymbol{p}/2 \; , \tag{1.96}$$

where $\boldsymbol{q}$ and $\boldsymbol{p}$ are independent two-dimensional vectors and rewrite (1.93) as

$$\psi(\boldsymbol{q} + \boldsymbol{p}/2, \boldsymbol{q} - \boldsymbol{p}/2) = \int d\boldsymbol{r} \exp(-i\boldsymbol{q} \cdot \boldsymbol{\rho})[\kappa_A(\boldsymbol{p}, \boldsymbol{q}; z)\delta\mu_a(\boldsymbol{r})$$
$$+ \kappa_D(\boldsymbol{p}, \boldsymbol{q}; z)\delta D(\boldsymbol{r})] \; , \tag{1.97}$$

The above result has the structure of a Fourier-Laplace transform by which $\delta\mu_a$ and δD are related to ψ. This relation can be used to obtain an inversion formula for the integral (1.93). To proceed, we note that the Fourier-transform in the transverse direction can be inverted separately from the Laplace transform in the longitudinal direction. We thus arrive at the result

$$\psi(\boldsymbol{q} + \boldsymbol{p}/2, \boldsymbol{q} - \boldsymbol{p}/2) = \int \left[\kappa_A(\boldsymbol{p}, \boldsymbol{q}; z)\tilde{\delta}\mu_a(\boldsymbol{q}, z) + \kappa_D(\boldsymbol{p}, \boldsymbol{q}; z)\tilde{\delta}D(\boldsymbol{q}, z) \right] dz \; , \tag{1.98}$$

where $\tilde{\delta}\mu_a$ and $\tilde{\delta}D$ denote the two-dimensional Fourier-transform. For fixed $\boldsymbol{q}$, (1.98) defines an integral equation for $\tilde{\delta}\mu_a(\boldsymbol{q}, z)$ and $\tilde{\delta}D(\boldsymbol{q}, z)$. It is readily seen that the minimum L^2 norm solution to (1.98) has the form

$$\tilde{\delta}\mu_a(\boldsymbol{q}, z) = \int dp dp' \kappa_A^*(\boldsymbol{p}, \boldsymbol{q}; z) M^{-1}(\boldsymbol{p}, \boldsymbol{p}'; \boldsymbol{q}) \psi(\boldsymbol{p}' + \boldsymbol{q}/2, \boldsymbol{p}' - \boldsymbol{q}/2) , \quad (1.99)$$

$$\tilde{\delta}D(\boldsymbol{q}, z) = \int dp dp' \kappa_D^*(\boldsymbol{p}, \boldsymbol{q}; z) M^{-1}(\boldsymbol{p}, \boldsymbol{p}'; \boldsymbol{q}) \psi(\boldsymbol{p}' + \boldsymbol{q}/2, \boldsymbol{p}' - \boldsymbol{q}/2) , \quad (1.100)$$

where the matrix elements of M are given by the integral

$$M(\boldsymbol{p}, \boldsymbol{p}'; \boldsymbol{q}) = \int_0^L [\kappa_A(\boldsymbol{p}, \boldsymbol{q}; z) \kappa_A^*(\boldsymbol{p}', \boldsymbol{q}; z) + \kappa_D(\boldsymbol{p}, \boldsymbol{q}; z) \kappa_D^*(\boldsymbol{p}', \boldsymbol{q}; z)] dz$$

$$(1.101)$$

Finally, we apply the inverse Fourier transform in the transverse direction to arrive at the inversion formula

$$\delta\mu_a(\boldsymbol{r}) = \int \frac{d\boldsymbol{q}}{(2\pi)^2} e^{-i\boldsymbol{q}\cdot\boldsymbol{\rho}}$$

$$\times \int dp dp' \kappa_A^*(\boldsymbol{p}, \boldsymbol{q}; z) M^{-1}(\boldsymbol{p}, \boldsymbol{p}'; \boldsymbol{q}) \psi(\boldsymbol{q}' + \boldsymbol{p}/2, \boldsymbol{q}' - \boldsymbol{p}/2) ,$$

$$(1.102)$$

$$\delta D(\boldsymbol{r}) = \int \frac{d\boldsymbol{q}}{(2\pi)^2} e^{-i\boldsymbol{q}\cdot\boldsymbol{\rho}}$$

$$\times \int dp dp' \kappa_D^*(\boldsymbol{p}, \boldsymbol{q}; z) M^{-1}(\boldsymbol{p}, \boldsymbol{p}'; \boldsymbol{q}) \psi(\boldsymbol{q}' + \boldsymbol{p}/2, \boldsymbol{q}' - \boldsymbol{p}/2) .$$

$$(1.103)$$

Several remarks on the above result are necessary. First, implementation of (1.102) and (1.103) requires regularization to stabilize the computation of the inverse of the matrix M. Second, sampling of the data function Φ_s is easily incorporated. The Fourier transform in (1.91) is replaced by a lattice Fourier transform. If the corresponding wave vectors $\boldsymbol{q}, \boldsymbol{q}_2$ are restricted to the first Brillouin zone of the lattice, then the inversion formula (1.102) and (1.103) recovers a bandlimited approximation to the coefficients $\delta\mu_a$ and δD [19, 20]. Third, the resolution of reconstructed images in the transverse and longitudinal directions is, in general, quite different. The transverse resolution is controlled by sampling and is determined by the highest spatial frequency that is present in the data. The longitudinal resolution is much lower due to the severe ill-posedness of the Laplace transform inversion which is implicit in (1.97), similar to the one-dimensional case discussed in Sect. 1.3.1. Finally, the inversion formula (1.102) and (1.103) can be used to develop a fast image reconstruction algorithm whose computational complexity scales as $O(NM \log M)$, where M is the number of detectors, N is the number of sources and $M \gg N$ [19]. The algorithm has recently been tested in noncontact optical tomography experiments. Quantitative reconstructions of complex phantoms with millimeter-scale features located centimeters within a highly-scattering medium have been reported [21, 22]. Data sets of order 10^8 source-detector pairs can be reconstructed in approximately 1 minute of CPU time on a 1.5 GHz computer. An example is shown in Fig. 1.3.

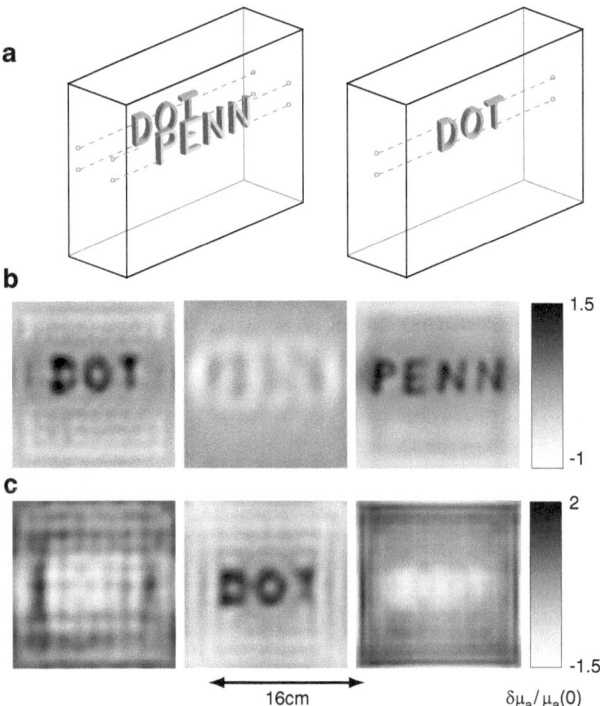

Fig. 1.3. Tomographic absorption images of a complex phantom using linear reconstructions within the diffusion approximation

Nonlinear Inverse Problem

We now consider the nonlinear inverse problem of DOT. The Born series (1.71) can be rewritten in the form

$$\Phi_s(\boldsymbol{r}_1, \boldsymbol{r}_2) = \int d\boldsymbol{r} K_1^i(\boldsymbol{r}_1, \boldsymbol{r}_2; \boldsymbol{r}) \eta_i(\boldsymbol{r})$$
$$+ \int d\boldsymbol{r} d\boldsymbol{r}' K_2^{ij}(\boldsymbol{r}_1, \boldsymbol{r}_2; \boldsymbol{r}, \boldsymbol{r}') \eta_i(\boldsymbol{r}) \eta_j(\boldsymbol{r}') + \cdots , \qquad (1.104)$$

where

$$\eta(\boldsymbol{r}) = \begin{pmatrix} \eta_1(\boldsymbol{r}) \\ \eta_2(\boldsymbol{r}) \end{pmatrix} = \begin{pmatrix} c\delta\mu_a(\boldsymbol{r}) \\ \delta D(\boldsymbol{r}) \end{pmatrix} , \qquad (1.105)$$

summation over repeated indices is implied with $i, j = 1, 2$. The components of the operators K_1 and K_2 are given by

$$K_1^1(\boldsymbol{r}_1, \boldsymbol{r}_2; \boldsymbol{r}) = G(\boldsymbol{r}_1, \boldsymbol{r})G(\boldsymbol{r}, \boldsymbol{r}_2) , \tag{1.106}$$

$$K_1^2(\boldsymbol{r}_1, \boldsymbol{r}_2; \boldsymbol{r}) = \nabla_{\boldsymbol{r}} G(\boldsymbol{r}_1, \boldsymbol{r}) \cdot \nabla_{\boldsymbol{r}} G(\boldsymbol{r}, \boldsymbol{r}_2) , \tag{1.107}$$

$$K_2^{11}(\boldsymbol{r}_1, \boldsymbol{r}_2; \boldsymbol{r}, \boldsymbol{r}') = -G(\boldsymbol{r}_1, \boldsymbol{r})G(\boldsymbol{r}, \boldsymbol{r}')G(\boldsymbol{r}', \boldsymbol{r}_2) , \tag{1.108}$$

$$K_2^{12}(\boldsymbol{r}_1, \boldsymbol{r}_2; \boldsymbol{r}, \boldsymbol{r}') = -G(\boldsymbol{r}_1, \boldsymbol{r})\nabla_{\boldsymbol{r}'} G(\boldsymbol{r}, \boldsymbol{r}') \cdot \nabla_{\boldsymbol{r}'} G(\boldsymbol{r}', \boldsymbol{r}_2) , \tag{1.109}$$

$$K_2^{21}(\boldsymbol{r}_1, \boldsymbol{r}_2; \boldsymbol{r}, \boldsymbol{r}') = -\nabla_{\boldsymbol{r}} G(\boldsymbol{r}_1, \boldsymbol{r}) \cdot \nabla_{\boldsymbol{r}} G(\boldsymbol{r}, \boldsymbol{r}')G(\boldsymbol{r}', \boldsymbol{r}_2) , \tag{1.110}$$

$$K_2^{22}(\boldsymbol{r}_1, \boldsymbol{r}_2; \boldsymbol{r}, \boldsymbol{r}') = -\nabla_{\boldsymbol{r}} G(\boldsymbol{r}_1, \boldsymbol{r}) \cdot \nabla_{\boldsymbol{r}} [\nabla_{\boldsymbol{r}'} G(\boldsymbol{r}, \boldsymbol{r}') \cdot$$
$$\times \nabla_{\boldsymbol{r}'} G(\boldsymbol{r}', \boldsymbol{r}_2)] . \tag{1.111}$$

It will prove useful to express the Born series as a formal power series in tensor powers of η of the form

$$\Phi_s = K_1\eta + K_2\eta \otimes \eta + K_3\eta \otimes \eta \otimes \eta + \cdots . \tag{1.112}$$

The solution to the nonlinear inverse problem of DOT may be formulated from the ansatz that η may be expressed as a series in tensor powers of Φ_s of the form

$$\eta = \mathcal{K}_1\Phi_s + \mathcal{K}_2\Phi_s \otimes \Phi_s + \mathcal{K}_3\Phi_s \otimes \Phi_s \otimes \Phi_s + \cdots , \tag{1.113}$$

where the $\mathcal{K}_j$'s are operators which are to be determined [23–26]. To proceed, we substitute the expression (1.112) for Φ_s into (1.113) and equate terms with the same tensor power of η. We thus obtain the relations

$$\mathcal{K}_1 K_1 = I , \tag{1.114}$$

$$\mathcal{K}_2 K_1 \otimes K_1 + \mathcal{K}_1 K_2 = 0 , \tag{1.115}$$

$$\mathcal{K}_3 K_1 \otimes K_1 \otimes K_1 + \mathcal{K}_2 K_1 \otimes K_2 + \mathcal{K}_2 K_2 \otimes K_1 + \mathcal{K}_1 K_3 = 0 , \tag{1.116}$$

$$\sum_{m=1}^{j-1} \mathcal{K}_m \sum_{i_1 + \cdots + i_m = j} K_{i_1} \otimes \cdots \otimes K_{i_m} + \mathcal{K}_j K_1 \otimes \cdots \otimes K_1 = 0 , \tag{1.117}$$

which may be solved for the $\mathcal{K}_j$'s with the result

$$\mathcal{K}_1 = K_1^+ , \tag{1.118}$$

$$\mathcal{K}_2 = -\mathcal{K}_1 K_2 \mathcal{K}_1 \otimes \mathcal{K}_1 , \tag{1.119}$$

$$\mathcal{K}_3 = -(\mathcal{K}_2 K_1 \otimes K_2 + \mathcal{K}_2 K_2 \otimes K_1 + \mathcal{K}_1 K_3) \mathcal{K}_1 \otimes \mathcal{K}_1 \otimes \mathcal{K}_1 , \tag{1.120}$$

$$\mathcal{K}_j = -\left(\sum_{m=1}^{j-1} \mathcal{K}_m \sum_{i_1 + \cdots + i_m = j} K_{i_1} \otimes \cdots \otimes K_{i_m} \right) \mathcal{K}_1 \otimes \cdots \otimes \mathcal{K}_1. \tag{1.121}$$

We will refer to (1.113) as the inverse series. Here we note several of its properties. First, K_1^+ is the regularized pseudoinverse of the operator K_1. The singular value decomposition of the operator K_1^+ can be computed analytically for particular geometries, as explained in sect. 1.3.1. Since the operator $\mathcal{K}_1$ is unbounded, regularization of K_1^+ is required to control the ill-posedness of the inverse problem. Second, the coefficients in the inverse series have a

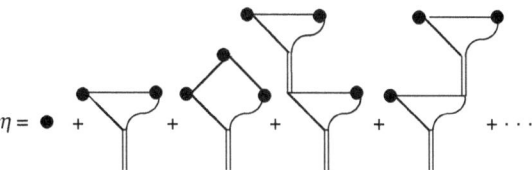

Fig. 1.4. Diagrammatic representation of the inverse scattering series

recursive structure. The operator $\mathcal{K}_j$ is determined by the coefficients of the Born series $K_1, K_2, \ldots, K_j$. Third, the inverse scattering series can be represented in diagrammatic form as shown in Fig. 1.4. A solid line corresponds to a factor of G, a wavy line to the incident field, a solid vertex ($\bullet$) to $\mathcal{K}_1\Phi_s$, and the effect of the final application of $\mathcal{K}_1$ is to join the ends of the diagrams. Note that the recursive structure of the series is evident in the diagrammatic expansion which is shown to third order. Finally, inversion of only the linear term in the Born series is required to compute the inverse series to all orders. Thus an ill-posed nonlinear inverse problem is reduced to an ill-posed linear inverse problem plus a well-posed nonlinear problem, namely the computation of the higher order terms in the series.

We now characterize the convergence of the inverse series. We restrict our attention to the case of a uniformly scattering medium for which $\eta = c\delta\mu_a$. To proceed, we require an estimate on the L^2 norm of the operator $\mathcal{K}_j$. We define the constants μ and ν by

$$\mu = \sup_{\boldsymbol{r} \in B_a} k^2 \|G_0(\boldsymbol{r}, \cdot)\|_{L^2(B_a)} . \tag{1.122}$$

$$\nu = k^2 |B_a|^{1/2} \sup_{\boldsymbol{r} \in B_a} \|G_0(\boldsymbol{r}, \cdot)\|_{L^2(\partial\Omega)} . \tag{1.123}$$

Here B_a denotes a ball of radius a which contains the support of η. It can be shown [26] that if $(\mu_p + \nu_p)\|\mathcal{K}_1\|_2 < 1$ then the operator

$$\mathcal{K}_j : L^2(\partial\Omega \times \cdots \times \partial\Omega) \longrightarrow L^2(B_a) \tag{1.124}$$

defined by (1.121) is bounded and

$$\|\mathcal{K}_j\|_2 \leq C(\mu + \nu)^j \|\mathcal{K}_1\|_2^j , \tag{1.125}$$

where $C = C(\mu, \nu, \|\mathcal{K}_1\|_2)$ is independent of j.

We can now state the main result on the convergence of the inverse series.

Theorem 1.1. *[26]. The inverse scattering series converges in the L^2 norm if $\|\mathcal{K}_1\|_2 < 1/(\mu+\nu)$ and $\|\mathcal{K}_1\Phi_s\|_{L^2(B_a)} < 1/(\mu+\nu)$. Furthermore, the following estimate for the series limit $\tilde{\eta}$ holds*

$$\left\| \tilde{\eta} - \sum_{j=1}^{N} \mathcal{K}_j \Phi_s \otimes \cdots \otimes \Phi_s \right\|_{L^2(B_a)} \leq C \frac{\left[(\mu_p + \nu_p)\|\mathcal{K}_1\Phi_s\|_{L^2(B_a)} \right]^{N+1}}{1 - (\mu + \nu)\|\mathcal{K}_1\Phi_s\|_{L^2(B_a)}} , \tag{1.126}$$

where $C = C(\mu, \nu, \|\mathcal{K}_1\|_2)$ does not depend on N nor on the scattering data Φ_s.

The stability of the limit of the inverse series under perturbations in the scattering data can be analyzed as follows:

Theorem 1.2. *[26]. Let $\|\mathcal{K}_1\|_2 < 1/(\mu + \nu)$ and let Φ_{s1} and Φ_{s2} be scattering data for which $M\|\mathcal{K}_1\|_2 < 1/(\mu + \nu)$, where $M = \max(\|\Phi_{s1}\|_2, \|\Phi_{s2}\|_2)$. Let η_1 and η_2 denote the corresponding limits of the inverse scattering series. Then the following estimate holds*

$$\|\eta_1 - \eta_2\|_{L^2(B_a)} < \tilde{C}\|\Phi_{s1} - \Phi_{s2}\|_{L^2(\partial\Omega \times \partial\Omega)} , \tag{1.127}$$

where $\tilde{C} = \tilde{C}(\mu, \nu, \|\mathcal{K}_1\|_2, M)$ is a constant that is otherwise independent of Φ_{s1} and Φ_{s2}.

It is a consequence of the proof of Theorem 1.2 that $\tilde{C}$ is proportional to $\|\mathcal{K}_1\|_2$. Since regularization sets the scale of $\|\mathcal{K}_1\|_2$, it follows that the stability of the nonlinear inverse problem is controlled by the stability of the linear inverse problem.

The limit of the inverse scattering series does not, in general, coincide with η. We characterize the approximation error as follows.

Theorem 1.3. *[26] Suppose that $\|\mathcal{K}_1\|_2 <, 1/(\mu + \nu)$, $\|\mathcal{K}_1\Phi_s\|_{L^2(B_a)} < 1/(\mu + \nu)$. Let $\mathcal{M} = \max(\|\eta\|_{L^2(B_a)}, \|\mathcal{K}_1 K_1 \eta\|_{L^2(B_a)})$ and assume that $\mathcal{M} < 1/(\mu + \nu)$. Then the norm of the difference between the partial sum of the inverse series and the true absorption obeys the estimate*

$$\left\|\eta - \sum_{j=1}^{N} \mathcal{K}_j \Phi_s \otimes \cdots \otimes \Phi_s\right\|_{L^2(B_a)} \leq C\|(I - \mathcal{K}_1 K_1)\eta\|_{L^2(B_a)}$$

$$+ \tilde{C}\frac{[(\mu_p + \nu_p)\|\mathcal{K}_1\|_2\|\Phi_s\|]^N}{1 - (\mu + \nu)\|\mathcal{K}_1\|_2\|\Phi_s\|_2} , \tag{1.128}$$

where $C = C(\mu, \nu, \|\mathcal{K}_1\|_2, \mathcal{M})$ and $\tilde{C} = \tilde{C}(\mu, \nu\|\mathcal{K}_1\|_2)$ are independent of N and Φ_s.

We note that, as expected, the above result shows that regularization of $\mathcal{K}_1$ creates an error in the reconstruction of η. For a fixed regularization, the relation $\mathcal{K}_1 K_1 = I$ holds on a subspace of $L^2(B_a)$ which, in practice, is finite dimensional. By regularizing $\mathcal{K}_1$ more weakly, the subspace becomes larger, eventually approaching all of L^2. However, in this instance, the estimate in Theorem 1.3 would not hold since $\|\mathcal{K}_1\|_2$ is so large that the inverse scattering series would not converge. Nevertheless, Theorem 1.3 does describe what can be reconstructed exactly, namely those η for which $\mathcal{K}_1 K_1 \eta = I$. That is, if we know apriori that η belongs to a particular finite-dimensional subspace of L^2, we can choose $\mathcal{K}_1$ to be a true inverse on this subspace. Then, if $\|\mathcal{K}_1\|_2$ and $\|\mathcal{K}_1\Phi_s\|_{L^2}$ are sufficiently small, the inverse series will recover η exactly.

It is straightforward to compute the constants μ and ν in dimension three. We have

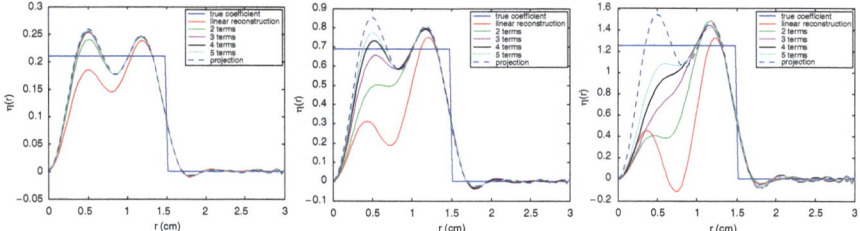

Fig. 1.5. Three-dimensional reconstructions of inhomogeneities with $R_1 = 1.5$ cm and $R_2 = 3$ cm. The contrast in absorption ranges from left to right: 1.2, 1.7 and 2.3

$$\mu = k^2 e^{-ka/2} \left(\frac{\sinh(ka)}{4\pi k} \right)^{1/2} . \tag{1.129}$$

$$\nu \le k^2 |\partial\Omega| |B_a|^{1/2} \frac{e^{-2k\,\mathrm{dist}(\partial\Omega, B_a)}}{(4\pi\,\mathrm{dist}(\partial\Omega, B_a))^2} . \tag{1.130}$$

Note that ν is exponentially small. It can be seen that the radius of convergence of the inverse series $R = 1/(\mu + \nu) \sim O(1/(ka)^{3/2})$ when $ka \gg 1$.

Numerical studies of the inverse scattering series have been performed. For inhomogeneities with radial symmetry, exact solutions to the forward problem were used as scattering data and reconstructions were computed to fifth order in the inverse series. It was found that the series appears to converge quite rapidly for low contrast objects. As the contrast is increased, the higher order terms systematically improve the reconstructions until, at sufficiently large contrast, the series diverges. See Fig. 1.5

1.3.2 Inverse Transport

We now turn our attention to the inverse problem for the RTE. This is a very large subject in its own right. A recent review has covered the key mathematical issues regarding existence, uniqueness and stability of the inverse transport problem [9]. Here we aim to discuss the inverse problem in some special cases which lead to direct inversion procedures. See [27] for further details. As in Sect. 1.3.1, we will work in the $z \ge 0$ half-space with the source and detector located on the $z = 0$ plane. The source is assumed to be pointlike and oriented in the inward normal direction. The light exiting the medium is further assumed to pass through a normally-oriented angularly-selective aperture which collects all photons with intensity

$$J(\boldsymbol{r}) = \int_{\hat{\nu}\cdot\hat{s}>0} \hat{\boldsymbol{\nu}} \cdot \hat{s} A(\hat{s}) I(\boldsymbol{r}, \hat{s}) d\hat{s} , \tag{1.131}$$

where A accounts for the effect of the aperture and the integration is carried out over all outgoing directions. When the aperture selects only photons traveling in the normal direction, then $A(\hat{s}) = \delta(\hat{s} - \hat{\nu})$ and $J(r) = I(r, \hat{\nu})$. This case is relevant to noncontact measurements in which the lens is focused at infinity. The case of complete angularly-averaged data corresponds to $A(\hat{s}) = 1$.

We now consider the linearized inverse problem. If the medium is inhomogeneously absorbing, it follows from (1.35) and (1.131) that the change in intensity measured relative to a homogeneous reference medium with absorption $\bar{\mu}_a$ is proportional to the data function $\Phi_s(\rho_1, \rho_2)$ which obeys the integral equation

$$\Phi_s(\rho_1, \rho_2) = \int_{\hat{\nu} \cdot \hat{s} > 0} \hat{\nu} \cdot \hat{s} A(\hat{s}) \Phi_s(\rho_1, 0, \hat{z}; \rho_2, 0, -\hat{s}) d\hat{s} . \tag{1.132}$$

Following the development in Sect. 1.3.1, we consider the Fourier transform of Φ_s with respect to the source and detector coordinates. Upon inserting the plane-wave decomposition for G given by (1.50) into (1.35) and carrying out the Fourier transform, we find that

$$\tilde{\Phi}_s(q_1, q_2) = \sum_{\mu_1, \mu_2} M_{\mu_1 \mu_2}(q_1, q_2) \int dr \exp\left[i(q_1 + q_2) \cdot \rho \right.$$
$$\left. - (Q_{\mu_1}(q_1) + Q_{\mu_2}(q_2)) z\right] \delta \mu_a(r) , \tag{1.133}$$

where

$$M_{\mu_1 \mu_2}(q_1, q_2) = \sum_{l_1 m_1, l_1' m_1'} \sum_{l_2 m_2} B_{l_1' m_1'}^{l_1 m_1}(q_1, \mu_1) B_{l_2 m_2}^{l_1' m_1'}(q_2, \mu_2)$$
$$\times \int_{\hat{\nu} \cdot \hat{s} > 0} \hat{\nu} \cdot \hat{s} A(\hat{s}) Y_{l_2 m_2}(\hat{s}) d\hat{s} . \tag{1.134}$$

Equation (1.133) is a generalization of the Fourier-Laplace transform which holds for the diffusion approximation, as discussed in Sect. 1.3.1. It can be seen that (1.133) reduces to the appropriate form in the diffuse limit, since only the smallest discrete eigenvalue contributes.

The inverse problem now consists of recovering $\delta \mu_a$ from $\tilde{\Phi}_s$. To proceed, we make use of the change of variables (1.96) and rewrite (1.133) as

$$\tilde{\Phi}_s(q + p/2, q - p/2) = \int dz K(q, p; z) \tilde{\delta \mu}_a(q, z) , \tag{1.135}$$

where $\tilde{\delta \mu}_a(q, z)$ denotes the two-dimensional Fourier transform of $\delta \mu_a$ with respect to its transverse argument and

$$K(q, p; z) = \sum_{\mu_1, \mu_2} M_{\mu_1 \mu_2}(q + p/2, q - p/2) \tag{1.136}$$
$$\times \exp\left[-(Q_{\mu_1}(q + p/2) + Q_{\mu_2}(q - p/2)) z\right] . \tag{1.137}$$

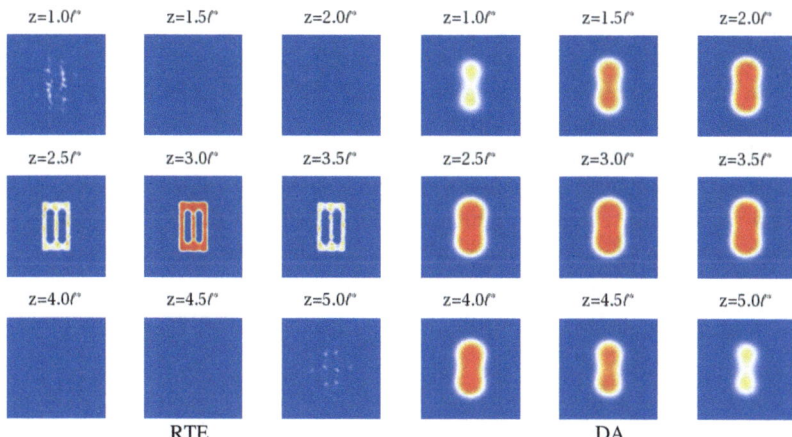

Fig. 1.6. Simulated linear reconstructions of $\delta\mu_a / \max(\delta\mu_a)$ for a bar target at various depths z. Results from the RTE and DA are compared. The field of view in each image is $2\,\mathrm{cm} \times 2\,\mathrm{cm}$ and the parameter $\ell^* = 1\,\mathrm{mm}$

This change of variables can be used to separately invert the transverse and longitudinal functional dependences of $\delta\mu_a$ since for fixed $\boldsymbol{q}$, (1.135) defines a one-dimensional integral equation for $\tilde{\delta\mu}_a(\boldsymbol{q}, z)$ whose pseudoinverse solution can in principle be computed. We thus obtain a solution to the inverse problem in the form

$$\delta\mu_a(\boldsymbol{r}) = \int \frac{d\boldsymbol{q}}{(2\pi)^2} e^{-i\boldsymbol{q}\cdot\boldsymbol{\rho}} \int d\boldsymbol{p} K^+(z; \boldsymbol{q}, \boldsymbol{p}) \tilde{\Phi}_s(\boldsymbol{q} + \boldsymbol{p}/2, \boldsymbol{q} - \boldsymbol{p}/2) \,, \quad (1.138)$$

where K^+ denotes the pseudoinverse of K. It is important to note the ill-posedness due to the exponential decay of the evanescent modes (1.47) for large z. Therefore, we expect that the resolution in the z direction will degrade with depth, but that sufficiently close to the surface the transverse resolution will be controlled by sampling. See Fig. 1.6 for the comparison of reconstructions of a bar target using the RTE and diffusion theory.

1.3.3 Single-Scattering Tomography

Consider an experiment in which a narrow collimated beam is normally incident on a highly-scattering medium which has the shape of a slab. Suppose that the slab is sufficiently thin that the incident beam undergoes predominantly single-scattering and that the intensity of transmitted light is measured by an angularly-selective detector. If the detector is collinear with the incident beam and its aperture is set to collect photons traveling in the normal direction, then only unscattered photons will be measured. Now, if the aperture is set away from the normal direction, then the detector will not register any photons. However, if the detector is no longer collinear and only photons which exit the slab at a fixed angle are collected, then it is possible to selectively measure single-scattered photons. Note that the contribution of single-scattered photons is described by (1.29).

The inverse problem for single-scattered light is to recover μ_a from measurements of δI as given by (1.29), assuming μ_s and p are known. The more general problem of simultaneously reconstructing μ_a, μ_s and p can also be considered. In either case, what must be investigated is the inversion of the broken-ray Radon transform which is defined as follows. Let f be a sufficiently smooth function. The broken-ray Radon transform is defined by

$$R_b f(\boldsymbol{r}_1, \hat{\mathbf{s}}_1; \boldsymbol{r}_2, \hat{\mathbf{s}}_2) = \int_{BR(\boldsymbol{r}_1, \hat{\mathbf{s}}_1; \boldsymbol{r}_2, \hat{\mathbf{s}}_2)} f(\boldsymbol{r}) d\boldsymbol{r} \ . \tag{1.139}$$

Here $BR(\boldsymbol{r}_1, \hat{\mathbf{s}}_1; \boldsymbol{r}_2, \hat{\mathbf{s}}_2)$ denotes the broken ray which begins at $\boldsymbol{r}_1$, travels in the direction $\hat{\mathbf{s}}_1$ and ends at $\boldsymbol{r}_2$ in the direction $\hat{\mathbf{s}}_2$. Note that if $\boldsymbol{r}_1, \boldsymbol{r}_2, \hat{\mathbf{s}}_1$ and $\hat{\mathbf{s}}_2$ all lie in the same plane and $\hat{\mathbf{s}}_1$ and $\hat{\mathbf{s}}_2$ point into and out of the slab, then the point of intersection $\boldsymbol{R}$ is uniquely determined. Thus it will suffice to consider the inverse problem in the plane and to reconstruct the function f from two-dimensional slices.

Evidently, the problem of inverting (1.139) is overdetermined. However, if the directions $\hat{\mathbf{s}}_1$ and $\hat{\mathbf{s}}_2$ are taken to be fixed, then the inverse problem is formally determined. To this end, we consider transmission measurements in a slab of width L and choose a coordinate system in which $\hat{\mathbf{s}}_1$ points in the $\hat{z}$ direction. The sources are chosen to be located on the line $z = 0$ in the yz-plane and are taken to point in the $\hat{z}$ direction. The detectors are located on the line $z = L$ and we assume that the angle θ between $\hat{\mathbf{s}}_1$ and $\hat{\mathbf{s}}_2$ is fixed. Under these conditions, it can be seen that the solution to (1.139) is given by

$$f(y, z) = \lambda \left\{ \left[\frac{\partial}{\partial \Delta} - (1 + \kappa) \frac{\partial}{\partial y} \right] \psi(y, \Delta) + \kappa \frac{\partial}{\partial y} \psi(y + \lambda z, \Delta_{\max}) \right. \\ \left. - \kappa(1 + \kappa) \frac{\partial^2}{\partial y^2} \int_\Delta^{\Delta_{\max}} \psi \left(y + \kappa(\ell - \Delta), \ell \right) d\ell \right\} \bigg|_{\Delta = (L - z)\tan\theta},$$

where $\Delta_{\max} = L\tan\theta$, $\lambda = \cot(\theta/2)$ and $\kappa = \cot(\theta/2)\cot\theta$.

Equation (1.140) is the inversion formula for the broken-ray Radon transform. We note that in contrast to x-ray computed tomography, it is unnecessary to collect projections along rays which are rotated about the sample. This considerably reduces the complexity of an imaging experiment. We also note that the presence of a derivative in (1.140), as in the Radon inversion formula, means that regularization is essential. Finally, by making use of measurements from multiple detector orientations it is possible to simultaneously reconstruct μ_a and μ_s. See [28] for further details.

We now illustrate the use of the inversion formula (1.140). Reconstructions are carried out in the rectangular area $0 \le z \le L$, $0 \le y \le 3L$ and the detection angle is is set to $\theta = \pi/4$, so that $\Delta_{\max} = L$. In what follows, $\delta\mu_t$ is decomposed as $\mu_t(y, z) = \bar{\mu}_t + \delta\mu_t(y, z)$, where $\bar{\mu}_t$ is the constant background

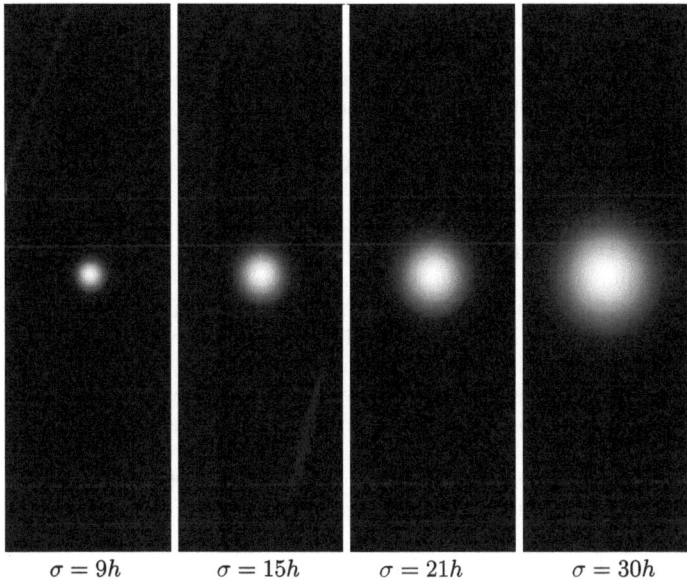

$$\sigma = 9h \qquad \sigma = 15h \qquad \sigma = 21h \qquad \sigma = 30h$$

Fig. 1.7. Reconstructed $\delta\mu_t(y, z)$ for the Gaussian inhomogeneity (1.140) using $N = 120$ and different values of σ, as labeled

value of the coefficient. We consider a smooth inhomogeneity in which the attenuation coefficient is a Gaussian function of the form

$$\delta\mu_t(y, z) = \bar{\mu}_t \exp\left[-\frac{(y - y_0)^2 + (z - z_0)^2}{\sigma^2} \right] . \tag{1.140}$$

The reconstructions are shown in Fig. 1.7.

Acknowledgments

This work was supported by the NSF under the grants DMS-0554100 and EEC-0615857, and by the NIH under the grant R01EB004832.

References

1. S.R. Arridge, Optical tomography in medical imaging. Inverse Probl. **15**(2), R41–R93 (1999)
2. M.W. van Rossum, T.W. Nieuwenhuizen, Multiple scattering of classical waves: microscopy, mesoscopy, and diffusion. Rev. Mod. Phys., **71**, 313–371 (1999)
3. A. Ishimaru, Wave Propagation and Scattering in Random Media. IEEE Press, Piscataway, NJ (1997)

4. K.M. Case, P.F. Zweifel, Linear transport theory. Addison-Wesley, Reading, MA, (1967)
5. C.L. Leakeas, E.W. Larsen, Generalized Fokker-Planck approximations of particle transport with highly forward-peaked scattering. Nucl. Sci. Eng. **137**, 236–250 (2001)
6. L. Ryzhik, G. Papanicolaou, J.B. Keller, Transport equations for elastic and other waves in random media. Wave Motion **24**, 327–370 (1996)
7. J. Sharpe et al., Optical projection tomography as a tool for 3D microscopy and gene expression studies. Science **296**, 541–545 (2002)
8. C. Vinegoni et al., In vivo imaging of Drosophila melanogaster pupae with mesoscopic fluorescence tomography. Nature Methods **5**, 45–47 (2008)
9. G. Bal, Inverse transport theory and applications. Inverse Probl. **25**(5), 053001 (48pp) (2009)
10. G. Bal, I. Langmore, F. Monard, Inverse transport with isotropic sources and angularly averaged measurements. Inverse Probl. Imag. **2**, 23–42 (2008)
11. G. Panasyuk, J.C. Schotland, V.A. Markel, Radiative transport equation in rotated reference frames. J. Phys. A **39**, 115–137 (2006)
12. A.D. Kim, J.B. Keller, Light propagation in biological tissue. J. Opt. Soc. Am A **20**, 92–98 (2003)
13. A.D. Kim, Transport theory for light propagation in biological tissue. J. Opt. Soc. Am A **21**, 820–827 (2004)
14. A.D. Kim, J.C. Schotland, Self-consistent scattering theory for the radiative transport equation. J. Opt. Soc. Am A **23**, 596–602 (2006)
15. E.W. Larsen, J.B. Keller, Asymptotic solution of neutron-transport problems for small mean free paths. J. Math. Phys. **15**, 75–81 (1974)
16. C.L. Epstein, J.C. Schotland, The bad truth about Laplace's transform. SIAM Rev. **50**, 504–520 (2008)
17. J.C. Schotland, Continuous-wave diffusion imaging. J. Opt. Soc. Am A **14**, 275–279 (1997)
18. J.C. Schotland, V.A. Markel, Inverse scattering with diffusing waves. J. Opt. Soc. Am A **18**, 2767–2777 (2001)
19. V.A. Markel, J.C. Schotland, Symmetries, inversion formulas, and image reconstruction for optical tomography. Phys. Rev. E **70** (2004)
20. V.A. Markel, J.C. Schotland, Effects of sampling and limited data in optical tomography. App. Phys. Lett. **81**, 1180–1182 (2002)
21. Z.M. Wang, G. Panasyuk, V.A. Markel, J.C. Schotland, Experimental demonstration of an analytic method for image reconstruction in optical diffusion tomography with large data sets. Opt. Lett. **30**, 3338–3340 (2005)
22. K. Lee, V. Markel, A.G. Yodh, S. Konecky, G.Y. Panasyuk, J.C. Schotland, Imaging complex structures with diffuse light. Opt. Express **16**, 5048–5060 (2008)
23. H.E. Moses, Calculation of the scattering potential from reflection coefficients. Phys. Rev. **102**, 550–67 (1956)
24. R. Prosser, Formal solutions of the inverse scattering problem. J. Math. Phys. **10**, 1819–1822 (1969)
25. V.A. Markel, J.A. O'Sullivan, J.C. Schotland, Inverse problem in optical diffusion tomography. IV. Nonlinear inversion formulas. J. Opt. Soc. Am A **20**, 903–912 (2003)
26. S. Moskow, J.C. Schotland, Convergence and stability of the inverse scattering series for diffuse waves. Inverse Probl. **24**, 065005 (16pp) (2008)

27. J.C. Schotland, V.A. Markel, Fourier-Laplace structure of the inverse scattering problem for the radiative transport equation. Inverse Probl. and Imag. **1**, 181–188 (2007)
28. V.A. Markel, L. Florescu, J.C. Schotland, Single-Scattering Optical Tomography. Phys. Rev. E **79** (2009)

2

Direct Reconstruction Methods in Ultrasound Imaging of Small Anomalies

Habib Ammari[1], Josselin Garnier[*,2], Vincent Jugnon[1], and Hyeonbae Kang[3]

[1] Department of Mathematics and Applications, Ecole Normale Supérieure, 45 Rue d'Ulm, 75005 Paris, France habib.ammari@ens.fr, vincent.jugnon@ens.fr
[2] Laboratoire de Probabilités et Modèles Aléatoires & Laboratoire Jacques-Louis Lions, Université Paris VII, Site Chevaleret, 75205 Paris Cedex 13, France garnier@math.jussieu.fr
[3] Department of Mathematics, Inha University, Incheon 402-751, Korea hbkang@inha.ac.kr

Summary. The aim of this chapter is to review direct (non-iterative) anomaly detection algorithms that take advantage of the smallness of the ultrasound anomalies. In particular, we numerically investigate their stability with respect to medium and measurement noises as well as their resolution.

2.1 Introduction

Ultrasound imaging is a noninvasive, easily portable, and relatively inexpensive diagnostic modality which finds extensive use in the clinic. The major clinical applications of ultrasound include many aspects of obstetrics and gynecology involving the assessment of fetal health, intra-abdominal imaging of the liver, kidney, and the detection of compromised blood flow in veins and arteries.

Operating typically at frequencies between 1 and 10 MHz, ultrasound imaging produces images via the backscattering of mechanical energy from interfaces between tissues and small structures within tissue. It has high spatial resolution, particularly at high frequencies, and involves no ionizing radiation. The weaknesses of the technique include the relatively poor soft-tissue contrast and the fact that gas and bone impede the passage of ultrasound waves, meaning that certain organs can not easily be imaged.

In this chapter, we review recent developments in the mathematical and numerical modelling of ultrasound anomaly detection at a fixed or multiple frequencies [1,3,6,7,9,11,12,20]. We construct different direct (non-iterative) anomaly detection algorithms that take advantage of the smallness of the ultrasound anomalies. In particular, MUltiple Signal Classification algorithm

H. Ammari (ed.), *Mathematical Modeling in Biomedical Imaging II*, Lecture Notes in Mathematics 2035, DOI 10.1007/978-3-642-22990-9_2,

(MUSIC), backpropagation, Kirchhoff migration, and topological derivative are investigated. Direct algorithms provide quite robust and accurate reconstruction of the location and of some geometric features of the anomalies, even with moderately noisy data. We numerically illustrate their stability with respect to medium and measurement noises as well as their resolution.

We also investigate multifrequency imaging. We illustrate that in the presence of (independent and identically distributed) measurement noises summing over frequencies a given imaging functional yields an improvement in the signal-to-noise ratio. However, if some correlation between frequency-dependent measurements exists, for example because of a medium noise, then summing over frequencies an imaging functional may not be appropriate. A single-frequency imaging functional at the frequency which maximizes the signal-to-noise ratio may give a better reconstruction.

When the acoustic medium is randomly heterogeneous, travel times cannot be known with accuracy so that images obtained with reverse time migration are noisy and not statistically stable, that is, they change with the realization of the random medium. Coherent Interferometry (CINT) has been shown to achieve a good compromise between resolution and deblurring for imaging in noisy environments from multiple frequency measurements [13,14]. CINT consists of backpropagating the cross correlations of the recorded signals over appropriate space-time or space-frequency windows rather than the signals themselves. We provide a CINT strategy in ultrasound imaging.

The chapter is organized as follows. In Sect. 2.2 we formulate a model problem and recall an asymptotic expansion of the boundary pressure perturbations due to small acoustic anomalies. In Sect. 2.3 we review direct imaging algorithms from measurements at fixed or multiple frequencies based on the asymptotic expansion. In Sect. 2.4 we perform a variety of numerical tests to compare the performance of the developed direct imaging algorithms in terms of resolution and stability with respect to measurement and medium noises.

2.2 Problem Formulation

Consider an acoustic anomaly with constant bulk modulus K and volumetric mass density ρ. The background medium $\Omega \subset \mathbb{R}^d$ is smooth and homogeneous with bulk modulus and density equal to one. Suppose that the operating frequency ω is such that ω^2 is not an eigenvalue for the operator $-\Delta$ in $L^2(\Omega)$ with homogeneous Neumann boundary conditions. The scalar acoustic pressure u generated by the Neumann data g in the presence of the anomaly D is the solution to the Helmholtz equation:

$$\begin{cases} \nabla \cdot \left(1_{\Omega\backslash\overline{D}}(\boldsymbol{x}) + \rho^{-1}1_D(\boldsymbol{x})\right)\nabla u + \omega^2\left(1_{\Omega\backslash\overline{D}}(\boldsymbol{x}) + K^{-1}1_D(\boldsymbol{x})\right)u = 0 & \text{in } \Omega, \\ \dfrac{\partial u}{\partial \nu} = g & \text{on } \partial\Omega, \end{cases}$$

$$(2.1)$$

while the background solution U satisfies

$$\begin{cases} \Delta U + \omega^2 U = 0 & \text{in } \Omega, \\ \dfrac{\partial U}{\partial \nu} = g & \text{on } \partial \Omega. \end{cases} \tag{2.2}$$

Here, ν is the outward normal to $\partial \Omega$ and $\mathbf{1}_D$ is the characteristic function of D.

The problem under consideration is the following one: given the field u measured at the surface of the domain Ω, we want to estimate the location of the anomaly D.

Suppose that the anomaly is $D = z_a + \delta B$, where z_a is the "center" of D, B is a smooth reference domain which contains the origin, and δ, the characteristic size of D, is a small parameter.

We provide an asymptotic expansion of the boundary pressure perturbations, $u - U$, as δ goes to zero. For doing so, we need to introduce a few auxiliary functions that can be computed either analytically or numerically.

For B a smooth bounded domain in $\mathbb{R}^d$ and $0 < k \neq 1 < +\infty$ a material parameter, let $\hat{v} = \hat{v}(k, B)$ be the solution to

$$\begin{cases} \Delta \hat{v} = 0 & \text{in } \mathbb{R}^d \setminus \overline{B}, \\ \Delta \hat{v} = 0 & \text{in } B, \\ \hat{v}|_- - \hat{v}|_+ = 0 & \text{on } \partial B, \\ k \dfrac{\partial \hat{v}}{\partial \nu}\bigg|_- - \dfrac{\partial \hat{v}}{\partial \nu}\bigg|_+ = 0 & \text{on } \partial B, \\ \hat{v}(\xi) - \xi \to 0 & \text{as } |\xi| \to +\infty. \end{cases} \tag{2.3}$$

Here we denote

$$v|_{\pm}(\xi) := \lim_{t \to 0^+} v(\xi \pm t\nu_\xi), \quad \xi \in \partial B,$$

and

$$\frac{\partial v}{\partial \nu}\bigg|_{\pm}(\xi) := \lim_{t \to 0^+} \nu_\xi^T \nabla v(\xi \pm t\nu_\xi), \quad \xi \in \partial B,$$

if the limits exist, where ν_ξ is the outward unit normal to ∂B at ξ and T stands for the transpose (so that $a^T b$ is the scalar product of the two vectors a and b). Recall that $\hat{v}$ plays the role of the first-order corrector in the theory of homogenization [18].

Define the polarization tensor $\mathbf{M}(k, B) = (M_{pq})_{p,q=1}^d$ by

$$M_{pq}(k, B) := (k-1) \int_B \frac{\partial \hat{v}_q}{\partial \xi_p}(\xi) \, d\xi, \tag{2.4}$$

where $\hat{v} = (\hat{v}_1, \ldots, \hat{v}_d)^T$ is the solution to (2.3). The formula of the polarization tensor for ellipses will be useful. In dimension $d = 2$, let B be an ellipse whose semi-axes are along the x_1- and x_2-axes and of length a and b, respectively. Then, $\mathbf{M}(k, B)$ takes the form [8]

$$\mathbf{M}(k, B) = (k - 1)|B| \begin{pmatrix} \dfrac{a+b}{a+kb} & 0 \\ 0 & \dfrac{a+b}{b+ka} \end{pmatrix}. \tag{2.5}$$

For $\omega \geq 0$, let for $\boldsymbol{x} \neq 0$,

$$\Gamma_\omega(\boldsymbol{x}) = \begin{cases} \dfrac{e^{i\omega|\boldsymbol{x}|}}{4\pi|\boldsymbol{x}|}, & d = 3, \\ \dfrac{i}{4} H_0^{(1)}(\omega|\boldsymbol{x}|), & d = 2, \end{cases} \tag{2.6}$$

which is the outgoing fundamental solution for the Helmholtz operator $-(\Delta + \omega^2)$ in $\mathbb{R}^d$. Here, $H_0^{(1)}$ is the Hankel function of the first kind of order zero.

Let the integral operator $\mathcal{K}_\Omega^\omega$ be defined by

$$\mathcal{K}_\Omega^\omega[\varphi](\boldsymbol{x}) = \int_{\partial\Omega} \frac{\partial \Gamma_\omega(\boldsymbol{x} - \boldsymbol{y})}{\partial \nu(\boldsymbol{y})} \varphi(\boldsymbol{y}) d\sigma(\boldsymbol{y}), \quad \boldsymbol{x} \in \partial\Omega.$$

For $\boldsymbol{z} \in \Omega$, let us now introduce the Neumann function for $-(\Delta + \omega^2)$ in Ω corresponding to a Dirac mass at $\boldsymbol{z}$. That is, N_ω is the solution to

$$\begin{cases} -(\Delta_x + \omega^2)N_\omega(\boldsymbol{x}, \boldsymbol{z}) = \delta_{\boldsymbol{z}} & \text{in } \Omega, \\ \dfrac{\partial N_\omega}{\partial \nu} = 0 & \text{on } \partial\Omega. \end{cases} \tag{2.7}$$

We will need the following lemma from [10, Proposition 2.8].

Lemma 2.1 *The following identity relating the fundamental solution Γ_ω to the Neumann function N_ω holds:*

$$\left(-\frac{1}{2}\mathcal{I} + \mathcal{K}_\Omega^\omega \right)[N_\omega(\cdot, \boldsymbol{z})](\boldsymbol{x}) = \Gamma_\omega(\boldsymbol{x} - \boldsymbol{z}), \quad \boldsymbol{x} \in \partial\Omega, \ \boldsymbol{z} \in \Omega. \tag{2.8}$$

In (2.8), $\mathcal{I}$ denotes the identity.

Assuming that ω^2 is not an eigenvalue for the operator $-\Delta$ in $L^2(\Omega)$ with homogeneous Neumann boundary conditions, we can prove, using the theory of relatively compact operators, the existence and uniqueness of a solution to (2.1) at least for δ small enough [19]. Moreover, the following asymptotic formula for boundary pressure perturbations that are due to the presence of a small acoustic anomaly holds [7, 19].

Theorem 2.2 *Let u be the solution of (2.1) and let U be the background solution. Suppose that $D = \boldsymbol{z}_a + \delta B$, with $0 < (K, \rho) \neq (1, 1) < +\infty$. Suppose that $\omega\delta \ll 1$.*

(i) *For any* $\boldsymbol{x} \in \partial\Omega$,

$$u(\boldsymbol{x}) = U(\boldsymbol{x}) - \delta^d \Big(\nabla U(\boldsymbol{z}_a)^T \mathbf{M}(1/\rho, B) \nabla_{\boldsymbol{z}} N_\omega(\boldsymbol{x}, \boldsymbol{z}_a)$$
$$+ \omega^2 (K^{-1} - 1)|B|U(\boldsymbol{z}_a) N_\omega(\boldsymbol{x}, \boldsymbol{z}_a) \Big) + o(\delta^d), \qquad (2.9)$$

where $\mathbf{M}(1/\rho, B)$ *is the polarization tensor associated with* B *and* $1/\rho$.

(ii) *Let* w *be a smooth function such that* $(\Delta + \omega^2)w = 0$ *in* Ω. *The weighted boundary measurements* $I_w[U, w]$ *defined by*

$$I_w[U, w] := \int_{\partial\Omega} (u - U)(\boldsymbol{x}) \frac{\partial w}{\partial \nu}(\boldsymbol{x}) \, d\sigma(\boldsymbol{x}) \qquad (2.10)$$

satisfies

$$I_w[U, w] = -\delta^d \Big(\nabla U(\boldsymbol{z}_a)^T \mathbf{M}(1/\rho, B) \nabla w(\boldsymbol{z}_a)$$
$$+ \omega^2 (K^{-1} - 1)|B|U(\boldsymbol{z}_a) w(\boldsymbol{z}_a) \Big) + o(\delta^d). \qquad (2.11)$$

The weighted boundary measurements $I_w[U, w]$ will be used for introducing MUSIC and backpropagation-type algorithms.

Suppose that the background domain contains P well-separated anomalies $D_p = \boldsymbol{z}_p + \delta B_p$, $p = 1, \ldots, P$, with volumetric mass density and bulk modulus denoted by ρ_p and K_p, respectively. Then, (2.9) yields

$$u(\boldsymbol{x}) = U(\boldsymbol{x}) - \delta^d \sum_{p=1}^{P} \Big(\nabla U(\boldsymbol{z}_p)^T \mathbf{M}(1/\rho_p, B_p) \nabla_{\boldsymbol{z}} N_\omega(\boldsymbol{x}, \boldsymbol{z}_p)$$
$$+ \omega^2 (K_p^{-1} - 1)|B_p|U(\boldsymbol{z}_p) N_\omega(\boldsymbol{x}, \boldsymbol{z}_p) \Big) + o(\delta^d)$$

for $\boldsymbol{x} \in \partial\Omega$.

Combining (2.9) and Lemma 2.1, the following corollary immediately holds.

Corollary 2.3 *For any* $\boldsymbol{x} \in \partial\Omega$,

$$\Big(-\frac{1}{2}\mathcal{I} + \mathcal{K}_\Omega^\omega \Big)[u - U](\boldsymbol{x}) = -\delta^d \Big(\nabla U(\boldsymbol{z}_a)^T \mathbf{M}(1/\rho, B) \nabla_{\boldsymbol{z}} \Gamma_\omega(\boldsymbol{x} - \boldsymbol{z}_a)$$
$$+ \omega^2 (K^{-1} - 1)|B|U(\boldsymbol{z}_a) \Gamma_\omega(\boldsymbol{x} - \boldsymbol{z}_a) \Big) + o(\delta^d). \qquad (2.12)$$

Note that the remainders in expansions (2.9), (2.11), and (2.12) are uniform with respect to $\boldsymbol{x} \in \Omega$ and hold only when the distance between $\boldsymbol{z}_a$ and $\partial\Omega$ is much larger than δ.

2.3 Direct Imaging Algorithms

2.3.1 Direct Imaging at a Fixed Frequency

In this section, we apply the asymptotic formulas (2.9) and (2.11) for the purpose of identifying the location and certain properties of the shape of the acoustic anomalies.

Consider P well-separated anomalies $D_p = z_p + \delta B_p$, $p = 1, \ldots, P$. The volumetric mass density and bulk modulus of D_p are denoted by ρ_p and K_p, respectively. Suppose that all the domains B_p are disks.

MUSIC-Type Algorithm

Let $(\boldsymbol{\theta}_1, \ldots, \boldsymbol{\theta}_n)$ be n unit vectors in $\mathbb{R}^d$. For $\boldsymbol{\theta} \in \{\boldsymbol{\theta}_1, \ldots, \boldsymbol{\theta}_n\}$, we assume that we are in possession of the boundary data u when the domain Ω is illuminated with the plane wave $U(\boldsymbol{x}) = e^{i\omega\boldsymbol{\theta}^T\boldsymbol{x}}$. Therefore, taking the harmonic function $w(\boldsymbol{x}) = e^{-i\omega\boldsymbol{\theta}'^T\boldsymbol{x}}$ for $\boldsymbol{\theta}' \in \{\boldsymbol{\theta}_1, \ldots, \boldsymbol{\theta}_n\}$ and using (2.5) shows that the weighted boundary measurement is approximately equal to

$$I_w[U, \omega] \simeq -\sum_{p=1}^{P} |D_p|\omega^2 \left(2\frac{\rho_p^{-1} - 1}{\rho_p^{-1} + 1}\boldsymbol{\theta}^T\boldsymbol{\theta}' + K_p^{-1} - 1\right)e^{i\omega(\boldsymbol{\theta}-\boldsymbol{\theta}')^T z_p}.$$

Define the response matrix $\mathbf{A} = (A_{ll'})_{l,l'=1}^n \in \mathbb{C}^{n\times n}$ by

$$A_{ll'} := I_{w_{l'}}[U_l, \omega], \tag{2.13}$$

where $U_l(\boldsymbol{x}) = e^{i\omega\boldsymbol{\theta}_l^T\boldsymbol{x}}$, $w_l(\boldsymbol{x}) = e^{-i\omega\boldsymbol{\theta}_l^T\boldsymbol{x}}$, $l = 1, \ldots, n$. It is approximately given by

$$A_{ll'} \simeq -\sum_{p=1}^{P} |D_p|\omega^2 \left(2\frac{\rho_p^{-1} - 1}{\rho_p^{-1} + 1}\boldsymbol{\theta}_l^T\boldsymbol{\theta}_{l'} + K_p^{-1} - 1\right)e^{i\omega(\boldsymbol{\theta}_l-\boldsymbol{\theta}_{l'})^T z_p},$$

for $l, l' = 1, \ldots, n$. Introduce the n-dimensional vector fields $\boldsymbol{g}^{(j)}$, defined for $j = 1, \ldots, d+1$, by

$$\boldsymbol{g}^{(j)}(\boldsymbol{x}) = \frac{1}{\sqrt{n}}\left(\boldsymbol{e}_j^T\boldsymbol{\theta}_1 e^{i\omega\boldsymbol{\theta}_1^T\boldsymbol{x}}, \ldots, \boldsymbol{e}_j^T\boldsymbol{\theta}_n e^{i\omega\boldsymbol{\theta}_n^T\boldsymbol{x}}\right)^T, \quad j = 1, \ldots, d, \tag{2.14}$$

and

$$\boldsymbol{g}^{(d+1)}(\boldsymbol{x}) = \frac{1}{\sqrt{n}}\left(e^{i\omega\boldsymbol{\theta}_1^T\boldsymbol{x}}, \ldots, e^{i\omega\boldsymbol{\theta}_n^T\boldsymbol{x}}\right)^T, \tag{2.15}$$

where $(\boldsymbol{e}_1, \ldots, \boldsymbol{e}_d)$ is an orthonormal basis of $\mathbb{R}^d$. Let $\mathbf{P}_{\text{noise}} = \mathbf{I} - \mathbf{P}$, where $\mathbf{P}$ is the orthogonal projection onto the range of $\mathbf{A}$. The MUSIC-type imaging functional is defined by

$$\mathcal{I}_{\mathrm{MU}}(z^S,\omega) := \left(\sum_{j=1}^{d+1} \| \mathbf{P}_{\mathrm{noise}} g^{(j)}(z^S) \|^2 \right)^{-1/2}. \tag{2.16}$$

This functional has large peaks only at the locations of the anomalies; see, e.g., [1].

Backpropagation-Type Algorithms

Let $(\boldsymbol{\theta}_1,\ldots,\boldsymbol{\theta}_n)$ be n unit vectors in $\mathbb{R}^d$. A backpropagation-type imaging functional at a single frequency ω is given by

$$\mathcal{I}_{\mathrm{BP}}(z^S,\omega) := \frac{1}{n} \sum_{l=1}^{n} e^{-2i\omega\boldsymbol{\theta}_l^T z^S} I_{w_l}[U_l,\omega], \tag{2.17}$$

where $U_l(\boldsymbol{x}) = w_l(\boldsymbol{x}) = e^{i\omega\boldsymbol{\theta}_l^T \boldsymbol{x}}$, $l = 1,\ldots,n$. Suppose that $(\boldsymbol{\theta}_1,\ldots,\boldsymbol{\theta}_n)$ are equidistant points on the unit sphere S^{d-1}. For sufficiently large n, we have

$$\frac{1}{n} \sum_{l=1}^{n} e^{i\omega\boldsymbol{\theta}_l^T \boldsymbol{x}} \simeq 4\left(\frac{\pi}{\omega}\right)^{d-2} \mathrm{Im}\{\Gamma_\omega(\boldsymbol{x})\} = \begin{cases} \mathrm{sinc}(\omega|\boldsymbol{x}|) & \text{for } d = 3, \\ J_0(\omega|\boldsymbol{x}|) & \text{for } d = 2, \end{cases} \tag{2.18}$$

where $\mathrm{sinc}(s) = \sin(s)/s$ is the sinc function and J_0 is the Bessel function of the first kind and of order zero.

Therefore, it follows that

$$\mathcal{I}_{\mathrm{BP}}(z^S,\omega) \simeq \sum_{p=1}^{P} |D_p|\omega^2 \left(2\frac{\rho_p^{-1}-1}{\rho_p^{-1}+1} - (K_p^{-1}-1) \right)$$
$$\times \begin{cases} \mathrm{sinc}(2\omega|z^S - z_p|) & \text{for } d = 3, \\ J_0(2\omega|z^S - z_p|) & \text{for } d = 2. \end{cases}$$

These formulae show that the resolution of the imaging functional is the standard diffraction limit. It is of the order of half the wavelength $\lambda = 2\pi/\omega$. Note that the above backpropagation-type algorithm is a simplified version of the algorithm studied in [4, 5]. In fact, instead of using only the diagonal terms of the response matrix $\mathbf{A}$, defined by (2.13), we can use the whole matrix to define the Kirchhoff migration functional:

$$\mathcal{I}_{\mathrm{KM}}(z^S,\omega) = \sum_{j=1}^{d+1} \overline{g^{(j)}(z^S)}^T \mathbf{A} g^{(j)}(z^S), \tag{2.19}$$

where $g^{(j)}$ are defined by (2.14) and (2.15).

Suppose for simplicity that $P = 1$ and $\rho = 1$. In this case the response matrix is

$$\mathbf{A} = -n|D|\omega^2(K^{-1} - 1)g^{(d+1)}(z_a)\overline{g^{(d+1)}(z_a)}^T$$

and we can prove that $\mathcal{I}_{\mathrm{MU}}$ is a nonlinear pointwise function of $\mathcal{I}_{\mathrm{KM}}$ [4]. In fact, we have

$$\mathcal{I}_{\mathrm{KM}}(\boldsymbol{z}^S, \omega) = -n|D|\omega^2(K^{-1}-1)\left(1 - \mathcal{I}_{\mathrm{MU}}^{-2}(\boldsymbol{z}^S, \omega)\right).$$

It is worth pointing out that this transformation does not improve neither the stability nor the resolution.

Moreover, in the presence of additive measurement noise, the response matrix can be written as

$$\mathbf{A} = -n|D|\omega^2(K^{-1}-1)\boldsymbol{g}^{(d+1)}(\boldsymbol{z}_a)\overline{\boldsymbol{g}^{(d+1)}(\boldsymbol{z}_a)}^T + \sigma\omega\mathbf{W},$$

where $\mathbf{W}$ is a complex symmetric Gaussian matrix with mean zero and variance 1. According to [4], the Signal-to-Noise Ratio (SNR) of the imaging functional $\mathcal{I}_{\mathrm{KM}}$, defined by

$$\mathrm{SNR}(\mathcal{I}_{\mathrm{KM}}) = \frac{\mathbb{E}[\mathcal{I}_{\mathrm{KM}}(\boldsymbol{z}_a, \omega)]}{\mathrm{Var}(\mathcal{I}_{\mathrm{KM}}(\boldsymbol{z}_a, \omega))^{1/2}},$$

is then equal to

$$\mathrm{SNR}(\mathcal{I}_{\mathrm{KM}}) = \frac{n\omega|D|\,|K^{-1}-1|}{\sigma}. \tag{2.20}$$

For the MUSIC algorithm, the peak of $\mathcal{I}_{\mathrm{MU}}$ is affected by measurement noise. We have [16]

$$\mathcal{I}_{\mathrm{MU}}(\boldsymbol{z}_a, \omega) = \begin{cases} \frac{n|D|\omega|K^{-1}-1|}{\sigma} & \text{if } n|D|\omega|K^{-1}-1| \gg \sigma, \\ 1 & \text{if } n|D|\omega|K^{-1}-1| \ll \sigma. \end{cases}$$

Suppose now that the medium is randomly heterogeneous around a constant background. Let K be the bulk modulus of the anomaly D. The index of refraction is of the form $1 + (K^{-1}-1)\mathbf{1}_D(\boldsymbol{x}) + \mu(\boldsymbol{x})$, where 1 stands for the constant background, $(K^{-1}-1)\mathbf{1}_D(\boldsymbol{x})$ stands for the localized perturbation of the index of refraction due to the anomaly, and $\mu(\boldsymbol{x})$ stands for the fluctuations of the index of refraction due to clutter (i.e., medium noise). We assume that μ is a random process with Gaussian statistics and mean zero, and that it is compactly supported within Ω.

If the random process μ has a small amplitude, then the background solution U, i.e., the field that would be observed without the anomaly, can be approximated by

$$U(\boldsymbol{x}) \simeq U^{(0)}(\boldsymbol{x}) - \omega^2 \int_\Omega N_\omega^{(0)}(\boldsymbol{x}, \boldsymbol{y})\mu(\boldsymbol{y})U^{(0)}(\boldsymbol{y})\,d\boldsymbol{y},$$

where $U^{(0)}$ and $N_\omega^{(0)}$ are respectively the background solution and the Neumann function in the constant background case. On the other hand, in the weak fluctuation regime, the phase mismatch between $N_\omega(x, z_a)$, the Neumann function in the random background, and $N_\omega^{(0)}(x, z^S)$ when z^S is close to z_a comes from the random fluctuations of the travel time between x and z_a which is approximately equal to the integral of $\mu/2$ along the ray from x to z_a:

$$N_\omega(x, z_a) \simeq N_\omega^{(0)}(x, z_a)e^{i\omega T(x)},$$

with

$$T(x) \simeq \frac{|x - z_a|}{2} \int_0^1 \mu\left(z_a + \frac{x - z_a}{|x - z_a|}s\right) ds.$$

Therefore, for any smooth function w satisfying $(\Delta + \omega^2)w = 0$ in Ω, the weighted boundary measurements $I_w[U^{(0)}, w]$, defined by (2.10), is approximately given by

$$\begin{aligned}
I_w[U^{(0)}, w] &\simeq -|D|\omega^2(K^{-1} - 1)e^{-\frac{\omega^2 \mathrm{Var}(T)}{2}} w(z_a)U^{(0)}(z_a) \\
&\quad -\omega^2 \int_\Omega w(y)U^{(0)}(y)\mu(y)\,dy,
\end{aligned} \tag{2.21}$$

provided that the correlation radius of the random process μ is small [3]. Expansion (2.21) shows that the medium noise reduces the height of the main peak of $\mathcal{I}_{KM}$ by the damping factor $e^{-\omega^2 \mathrm{Var}(T)/2}$ and on the other hand it induces random fluctuations of the associated image in the form of a speckle field.

Topological Derivative Based Imaging Functional

The topological derivative based imaging functional was introduced in [3].

Let $D' = z^S + \delta'B'$, $K' > 1$, $\rho' > 1$, B' be chosen a priori, and let δ' be small. If $K < 1$ and $\rho < 1$, then we choose $K' < 1$ and $\rho' < 1$.

Let w be the solution of the Helmholtz equation

$$\begin{cases}
\Delta w + \omega^2 w = 0 & \text{in } \Omega, \\
\dfrac{\partial w}{\partial \nu} = (-\tfrac{1}{2}\mathcal{I} + (\mathcal{K}_\Omega^{-\omega})^*)\overline{(-\tfrac{1}{2}\mathcal{I} + \mathcal{K}_\Omega^\omega)[U - u_{\mathrm{meas}}]} & \text{on } \partial\Omega,
\end{cases} \tag{2.22}$$

where u_{meas} is the boundary pressure in the presence of the acoustic anomaly. The function w is obtained by backpropagating the Neumann data

$$(-\tfrac{1}{2}\mathcal{I} + (\mathcal{K}_\Omega^\omega)^*)\overline{(-\tfrac{1}{2}\mathcal{I} + \mathcal{K}_\Omega^\omega)[U - u_{\mathrm{meas}}]}$$

inside the background medium (without any anomaly). Note that $\overline{(\mathcal{K}_\Omega^\omega)^*} = (\mathcal{K}_\Omega^{-\omega})^*$.

The function w can be used to image the anomaly. It corresponds to backpropagating the discrepancy between the measured and the background solutions. However, we introduce here a functional that exploits better the coherence between the phases of the background and perturbed fields at the location of the anomaly. This functional turns out to be exactly the topological derivative imaging functional introduced [3].

For a single measurement, we set

$$
\mathcal{I}_{\mathrm{TD}}[U](z^S) = \mathrm{Re}\Big\{ \nabla U(z^S)^T \mathbf{M}(1/\rho', B') \nabla w(z^S)
$$

$$
+ \omega^2 (K'^{-1} - 1)|B'| U(z^S) w(z^S) \Big\}. \tag{2.23}
$$

The functional $\mathcal{I}_{\mathrm{TD}}[U](z^S)$ gives, at every search point $z^S \in \Omega$, the sensitivity of the misfit function

$$
\mathcal{E}[U](z^S) := \frac{1}{2} \int_{\partial\Omega} \Big|(-\frac{1}{2}\mathcal{I} + \mathcal{K}_\Omega^\omega)[u_{z^S} - u_{\mathrm{meas}}](x)\Big|^2 d\sigma(x),
$$

where u_{z^S} is the solution of (2.1) with the anomaly $D' = z^S + \delta' B'$. The location of the maximum of $z^S \mapsto \mathcal{I}_{\mathrm{TD}}[U](z^S)$ corresponds to the point at which the insertion of an anomaly centered at that point maximally decreases the misfit function. Using Corollary 2.3, we can show that the functional $\mathcal{I}_{\mathrm{TD}}$ attains its maximum at $z^S = z_a$; see [3]. It is also shown in [3] that the postprocessing of the data set by applying the integral operator $(-\frac{1}{2}\mathcal{I} + \mathcal{K}_\Omega^\omega)$ is essential in order to obtain an efficient topological based imaging functional, both in terms of resolution and stability. By postprocessing the data, we ensure that the topological based imaging functional attains its maximum at the true location of the anomaly.

For multiple measurements, $U_l, l = 1, \ldots, n$, the topological derivative based imaging functional is simply given by

$$
\mathcal{I}_{\mathrm{TD}}(z^S, \omega) := \frac{1}{n} \sum_{l=1}^{n} \mathcal{I}_{\mathrm{TD}}[U_l](z^S). \tag{2.24}
$$

Let, for simplicity, $(\theta_1, \ldots, \theta_n)$ be n uniformly distributed directions over the unit sphere and consider U_l to be the plane wave

$$
U_l(x) = e^{i\omega\theta_l^T x}, \quad x \in \Omega, \quad l = 1, \ldots, n. \tag{2.25}
$$

Let

$$
r_\omega(z^S, z) := \int_{\partial\Omega} \Gamma_\omega(x - z^S)\overline{\Gamma_\omega}(x - z)\, d\sigma(x), \tag{2.26}
$$

$$
\mathbf{R}_\omega(z^S, z) := \int_{\partial\Omega} \nabla_z \Gamma_\omega(x - z^S)\nabla_z \overline{\Gamma_\omega}(x - z)^T\, d\sigma(x). \tag{2.27}
$$

Note that $\mathbf{R}_\omega(\boldsymbol{z}^S, \boldsymbol{z})$ is a $d \times d$ matrix. When $\rho = 1$, we have

$$\mathcal{I}_{\mathrm{TD}}[U](\boldsymbol{z}^S) \simeq \delta^d \omega^4 (K'^{-1} - 1)(K^{-1} - 1)|B'|\mathrm{Re}\Big\{U(\boldsymbol{z}^S)r_\omega(\boldsymbol{z}^S, \boldsymbol{z}_a)\overline{U}(\boldsymbol{z}_a)\Big\},$$
$$(2.28)$$

where r_ω is given by (2.26). Therefore, by computing the topological derivatives for the n plane waves (n sufficiently large), it follows from (2.18) together with

$$\int_{\partial\Omega} \overline{\Gamma_\omega}(\boldsymbol{x} - \boldsymbol{z})\Gamma_\omega(\boldsymbol{x} - \boldsymbol{z}^S)\, d\sigma(\boldsymbol{x}) \sim \frac{1}{\omega}\mathrm{Im}\{\Gamma_\omega(\boldsymbol{z}^S - \boldsymbol{z})\}, \qquad d = 2, 3, \quad (2.29)$$

where $A \sim B$ means $A \simeq CB$ for some constant C independent of ω, that

$$\frac{1}{n}\sum_{l=1}^n \mathcal{I}_{\mathrm{TD}}[U_l](\boldsymbol{z}^S) \sim \omega^{5-d}(\mathrm{Im}\{\Gamma_\omega(\boldsymbol{z}^S - \boldsymbol{z}_a)\})^2.$$

Similarly, when $K = 1$, by computing the topological derivatives for the n plane waves, $U_l, l = 1, \ldots, n$, given by (2.25), we obtain

$$\frac{1}{n}\sum_{l=1}^n \mathcal{I}_{\mathrm{TD}}[U_l](\boldsymbol{z}^S)$$

$$\simeq \delta^d \omega^2 \frac{1}{n}\sum_{l=1}^n \mathrm{Re}\Big\{e^{i\omega\boldsymbol{\theta}_l^T(\boldsymbol{z}^S - \boldsymbol{z}_a)}\big[\boldsymbol{\theta}_l^T \mathbf{M}(1/\rho', B')\mathbf{R}_\omega(\boldsymbol{z}^S, \boldsymbol{z}_a)\mathbf{M}(1/\rho, B)^T \boldsymbol{\theta}_l\big]\Big\}.$$

Using $\rho' = 0$ and B' the unit disk, the polarization tensor $\mathbf{M}(1/\rho', B') = C_d\mathbf{I}$, where C_d is a constant, is proportional to the identity [8]. If, additionally, we assume that $\mathbf{M}(1/\rho, B)$ is approximately proportional to the identity, which occurs in particular when B is a disk or a ball, then by using

$$\int_{\partial\Omega} \nabla_{\boldsymbol{z}}\Gamma_\omega(\boldsymbol{x} - \boldsymbol{z}^S)\nabla_{\boldsymbol{z}}\overline{\Gamma_\omega}(\boldsymbol{x} - \boldsymbol{z})^T \, d\sigma(\boldsymbol{x})$$

$$\sim \omega\,\mathrm{Im}\{\Gamma_\omega(\boldsymbol{z}^S - \boldsymbol{z})\}\left(\frac{\boldsymbol{z} - \boldsymbol{z}^S}{|\boldsymbol{z} - \boldsymbol{z}^S|}\right)\left(\frac{\boldsymbol{z} - \boldsymbol{z}^S}{|\boldsymbol{z} - \boldsymbol{z}^S|}\right)^T, \qquad (2.30)$$

we arrive at

$$\frac{1}{n}\sum_{l=1}^n \mathcal{I}_{\mathrm{TD}}[U_l](\boldsymbol{z}^S) \sim \omega^{5-d}(\mathrm{Im}\{\Gamma_\omega(\boldsymbol{z}^S - \boldsymbol{z}_a)\})^2. \qquad (2.31)$$

Therefore, $\mathcal{I}_{\mathrm{TD}}$ attains its maximum at $\boldsymbol{z}_a$. Moreover, the resolution for the location estimation is given by the diffraction limit. We refer the reader to [3] for a detailed stability analysis of $\mathcal{I}_{\mathrm{TD}}$ with respect to both medium and measurement noises as well as its resolution. In the case of measurement noise, the SNR of $\mathcal{I}_{\mathrm{TD}}$,

$$\text{SNR}(\mathcal{I}_{\text{TD}}) = \frac{\mathbb{E}[\mathcal{I}_{\text{TD}}(z_a, \omega)]}{\text{Var}(\mathcal{I}_{\text{TD}}(z_a, \omega))^{1/2}},$$

is equal to

$$\text{SNR}(\mathcal{I}_{\text{TD}}) = \frac{\sqrt{2}\pi^{1-d/2}\omega^{(d+1)/2}|U(z_a)|(K^{-1}-1)|D|}{\sigma},$$

where σ^2 is the noise variance.

In the case of medium noise, let us introduce the kernel

$$Q(z^S, z_a) := \text{Re}\left\{ U^{(0)}(z^S)\overline{U^{(0)}(z_a)} \int_{\partial\Omega} \Gamma_\omega(x - z^S)\overline{\Gamma_\omega(x - z_a)} \, d\sigma(x) \right\}.$$

We can express the topological derivative imaging functional as follows [3]:

$$\mathcal{I}_{\text{TD}}[U^{(0)}](z^S) \simeq \omega^4(K'^{-1} - 1)|B'| \int_\Omega \mu(y)Q(z^S, y) \, dy$$
$$+ \omega^4(K'^{-1} - 1)(K^{-1} - 1)|B'||D|Q(z^S, z_a)e^{-\frac{\omega^2 \text{Var}(T)}{2}},$$
$$\tag{2.32}$$

provided, once again, that the correlation radius of the random process μ is small. Consequently, the topological derivative has the form of a peak centered at the location z_a of the anomaly (second term of the right-hand side of (2.32)) buried in a zero-mean Gaussian field or speckle pattern (first term of the right-hand side of (2.32)) that we can characterize statistically.

2.3.2 Direct Imaging at Multiple Frequencies

Let $(\theta_1, \ldots, \theta_n)$ be n uniformly distributed directions over the unit sphere. We consider plane wave illuminations at multiple frequencies, $(\omega_k)_{k=1,\ldots,m}$, instead of a fixed frequency:

$$U_{lk}(x) := U(x, \theta_l, \omega_k) = e^{i\omega_k \theta_l^T x},$$

and record the acoustic perturbations due to the anomaly. In this case, we can construct the topological derivative imaging functional by summing over frequencies

$$\mathcal{I}_{\text{TDF}}(z^S) := \frac{1}{m} \sum_{k=1}^m \mathcal{I}_{\text{TD}}(z^S, \omega_k). \tag{2.33}$$

Suppose for simplicity that $\rho = 1$. Then, (2.28) and (2.29) yield

$$\mathcal{I}_{\text{TDF}}(z^S) \sim \int_\omega \omega^{5-d}\left(\text{Im}\{\Gamma_\omega(z^S - z_a)\} \right)^2 d\omega, \quad d = 2, 3,$$

and hence, $\mathcal{I}_{\text{TDF}}(z^S)$ has a large peak only at z_a. In the general case, we can use (2.30) to state the same behavior at z_a.

An alternative imaging functional when searching for an anomaly using multiple frequencies is the reverse time migration imaging functional [11]:

$$\mathcal{I}_{\mathrm{RMF}}(\boldsymbol{z}^S) := \frac{1}{nm} \sum_{l=1}^{n} \sum_{k=1}^{m} \overline{U}(\boldsymbol{z}^S, \boldsymbol{\theta}_l, \omega_k)$$
$$\times \int_{\partial\Omega} (-\frac{\mathcal{I}}{2} + \mathcal{K}_{\Omega}^{\omega_k})(u - U)(\boldsymbol{x}, \boldsymbol{\theta}_l, \omega_k) \overline{\Gamma_{\omega_k}}(\boldsymbol{x}, \boldsymbol{z}^S) \, d\sigma(\boldsymbol{x}).$$

(2.34)

In fact, when for instance $\rho = 1$,

$$\mathcal{I}_{\mathrm{RMF}}(\boldsymbol{z}^S) \sim \frac{1}{nm} \sum_{l=1}^{m} \sum_{k=1}^{m} \omega_k^3 U(\boldsymbol{z}_a, \boldsymbol{\theta}_l, \omega_k) \overline{U}(\boldsymbol{z}^S, \boldsymbol{\theta}_l, \omega_k) \mathrm{Im}\{\Gamma_{\omega_k}(\boldsymbol{z}^S - \boldsymbol{z}_a)\},$$

and therefore, it is approximately proportional to

$$\int_{\mathcal{S}^{d-1}} \int_{\omega} \omega^3 e^{i\omega\boldsymbol{\theta}^T(\boldsymbol{z}^S - \boldsymbol{z}_a)} \mathrm{Im}\{\Gamma_\omega(\boldsymbol{z}^S - \boldsymbol{z}_a)\} d\omega d\sigma(\boldsymbol{\theta})$$
$$\sim \int_{\omega} \omega^{5-d} \left(\mathrm{Im}\{\Gamma_\omega(\boldsymbol{z}^S - \boldsymbol{z}_a)\} \right)^2 d\omega,$$

where $\mathcal{S}^{d-1}$ is the unit sphere and $d = 2, 3$. Hence,

$$\mathcal{I}_{\mathrm{RMF}}(\boldsymbol{z}^S) \sim \mathcal{I}_{\mathrm{TDF}}(\boldsymbol{z}^S).$$

Finally, it is also possible to use a backpropagation imaging functional:

$$\mathcal{I}_{\mathrm{BPF}}(\boldsymbol{z}^S) := \frac{1}{m} \sum_{k=1}^{m} \mathcal{I}_{\mathrm{BP}}(\boldsymbol{z}^S, \omega),$$

or a Kirchhoff imaging functional:

$$\mathcal{I}_{\mathrm{KMF}}(\boldsymbol{z}^S) := \frac{1}{m} \sum_{k=1}^{m} \mathcal{I}_{\mathrm{KM}}(\boldsymbol{z}^S, \omega).$$

We contrast this with the matched field imaging functional:

$$\mathcal{I}_{\mathrm{MF}}(\boldsymbol{z}^S) := \frac{1}{m} \sum_{k=1}^{m} |\mathcal{I}_{\mathrm{KM}}(\boldsymbol{z}^S, \omega)|^2,$$

in which the phase coherence between the different frequency-dependent acoustic perturbations is not exploited. This makes sense when the different frequency-dependent perturbations are incoherent.

If the measurement noises $\nu_{\mathrm{noise}}(\boldsymbol{x}, \omega_k)$, $k = 1, \ldots, m$, are independent and identically distributed, the multiple frequencies enhance the detection performance via a higher "effective" SNR.

If some correlation between frequency-dependent perturbations exist, for example because of a medium noise, then summing over frequencies an imaging functional is not appropriate. A single-frequency imaging functional at the frequency which maximizes the SNR may give a better reconstruction.

In the presence of a medium noise, a CINT procedure may be appropriate. Following [13, 14] a CINT-like algorithm is given by

$$
\mathcal{I}_{\mathrm{CINT}}(z^S) = \int_{S^{d-1}} \int_{\omega_1} \int_{\omega_2} \int_{\partial\Omega} \int_{\partial\Omega} e^{-\frac{|\omega_1-\omega_2|^2}{2\Omega_D^2}} e^{-\frac{|x_1-x_2|^2}{2X_D^2}}
$$
$$
\times (-\frac{\mathcal{I}}{2} + \mathcal{K}_\Omega^{\omega_1})(u-U)(x_1,\boldsymbol{\theta},\omega_1)\overline{\Gamma_{\omega_1}}(x_1,z^S)\overline{U}(z^S,\boldsymbol{\theta},\omega_1)
$$
$$
\times \overline{(-\frac{\mathcal{I}}{2} + \mathcal{K}_\Omega^{\omega_2})(u-U)(x_2,\boldsymbol{\theta},\omega_2)\Gamma_{\omega_2}(x_2,z^S)}
$$
$$
\times U(z^S,\boldsymbol{\theta},\omega_2)d\sigma(x)d\omega_1 d\omega_2 d\sigma(\boldsymbol{\theta}), \tag{2.35}
$$

where X_D and Ω_D are two cut-off parameters.

The purpose of the CINT-like imaging functional $\mathcal{I}_{\mathrm{CINT}}$ is to keep in (2.35) the pairs (x_1,ω_1) and (x_2,ω_2) for which the postprocessed data $(-\frac{1}{2}\mathcal{I} + \mathcal{K}_\Omega^{\omega_1})[u-U](x_1,\omega_1)$ and $(-\frac{1}{2}\mathcal{I} + \mathcal{K}_\Omega^{\omega_2})[u-U](x_1,\omega_1)$ are coherent, and to remove the pairs that do not bring information.

Depending on the parameters X_D, Ω_D, we get different trade-offs between resolution and stability. When X_D and Ω_D become small, $\mathcal{I}_{\mathrm{CINT}}$ presents better stability properties at the expense of a loss of resolution. In the limit $X_D \to \infty$, $\Omega_D \to \infty$, we get the square of the topological derivative functional $\mathcal{I}_{\mathrm{TDF}}$. A precise stability and resolution analysis for $\mathcal{I}_{\mathrm{CINT}}$ can be derived by exactly the same arguments as those in [2].

2.4 Numerical Illustrations

In this section we present results of numerical experiments to illustrate the performance of the imaging functionals introduced in the previous section.

We consider the two-dimensional case $(d = 2)$. The domain Ω is the unit disk. We simulate the measurements using a finite-element method to solve the Helmholtz equation. We use a piecewise linear representation of the solution u and piecewise constant representations of the parameter distributions $\mathbf{1}_{\Omega\backslash\overline{D}}(x)+\rho^{-1}\mathbf{1}_D(x)$ and $\mathbf{1}_{\Omega\backslash\overline{D}}(x)+K^{-1}\mathbf{1}_D(x)$. We consider a small anomaly $D = z_a + \delta B$ with $z_a = (-0.3, 0.5)$, $\delta = 0.05$, and B being the unit disk.

2.4.1 Measurements at a Fixed Frequency

We fix the working frequency ω to be equal to 6, which corresponds to a wavelength $\lambda \simeq 1$. We assume that the measurements correspond to the plane wave illuminations, $U_l(x) = e^{i\omega\boldsymbol{\theta}_l^T x}$, at the equi-distributed directions $\boldsymbol{\theta}_l$, for $l = 1,\ldots,n = 50$.

Bulk Modulus Contrast Only

Here, the parameters of the anomaly are $\rho = 1$ and $K = 1/2$.

Resolution in the Absence of Noise

Within the above setting, we first present results of the described algorithms in the absence of noise. In Fig. 2.1, plots of $\mathcal{I}_{\mathrm{TD}}(z^S, \omega)$, defined by (2.24), with $n = 50$ and $n = 2$ illustrate the efficiency of the proposed topological derivative based imaging procedure. The imaging functional $\mathcal{I}_{\mathrm{TD}}(z^S, \omega)$ reaches its maximum at the location z_a of the anomaly and behaves, accordingly to (2.30), like $J_0(\omega|z^S - z_a|)^2$ if the number n of incident waves is large while for small n, it behaves, as expected, like $J_0(\omega|z^S - z_a|)$.

In Fig. 2.2, we present two MUSIC-type reconstructions. Given the structure of the response matrix $\mathbf{A}$ with $\rho = 1$ (contrast only on the K distribution), it is known that its SVD yields only one significant singular value. See, *e.g.*, [1,6]. Thus, the illumination vectors $g^{(1)}$ and $g^{(2)}$ (see (2.14)) do not belong to the signal subspace of $\mathbf{A}$. Using these vectors in the projection step generates a blurred MUSIC image (figure on the left). To get a sharp peak, we should project only the illumination vector $g^{(3)}$ (figure on the right), which assumes a priori knowledge of the physical nature of the contrast.

As shown in Fig. 2.3, the backpropagation image of the anomaly has the expected behavior of the Bessel function and reaches its maximum at the location of the anomaly.

Stability with Respect to Measurement and Medium Noises

We now consider imaging at a fixed frequency from noisy data. We first add electronic (measurement) noise ν_{noise} to the previous measurements $u_{i,\mathrm{meas}}$, $i = 1, \ldots, n$. Here, ν_{noise} is a white Gaussian noise with standard

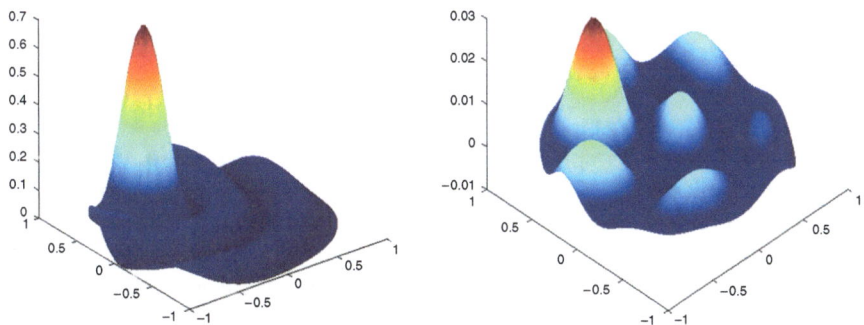

Fig. 2.1. Plots of $\mathcal{I}_{\mathrm{TD}}(z^S, \omega)$ defined by (2.24) with $n = 50$ (*left*) and $n = 2$ (*right*)

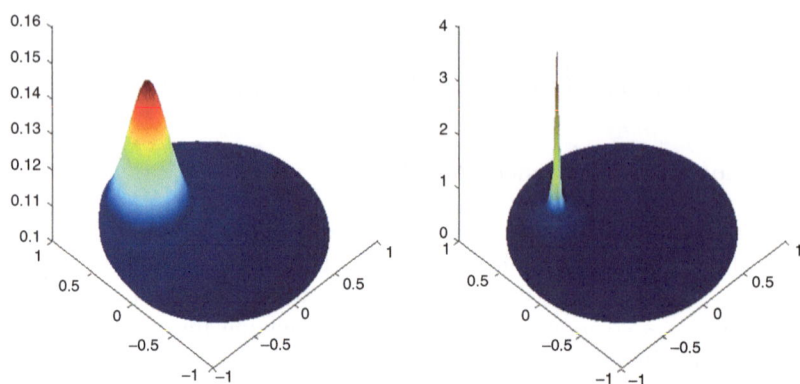

Fig. 2.2. *Left*: MUSIC image using the projection of $g^{(1)}, g^{(2)}$, and $g^{(3)}$ on the signal subspace of $\mathbf{A}$. *Right*: MUSIC image using the projection of $g^{(3)}$ on the signal subspace of $\mathbf{A}$

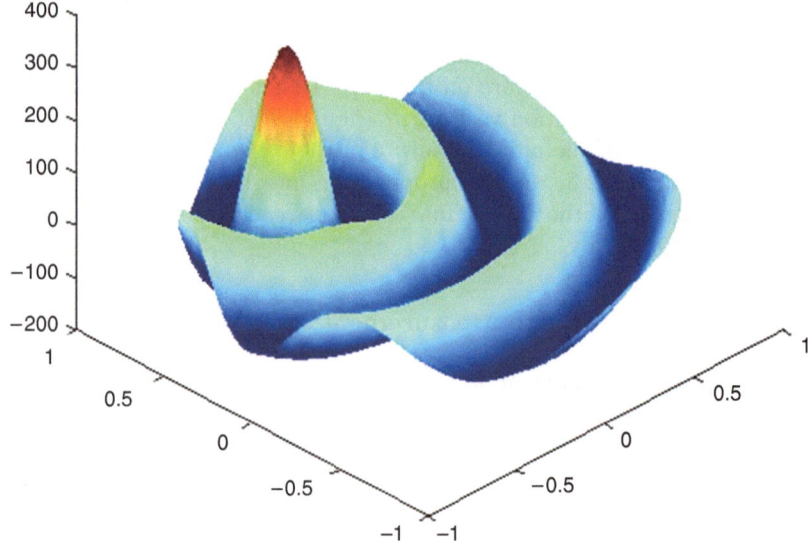

Fig. 2.3. Plot of $\mathcal{I}_{\mathrm{BP}}(z^S, \omega)$ defined by (2.17) with $n = 50$

deviation $\sigma\%$ of the L^2 norm of u_{meas} and σ ranges from 0 to 30. We compute $N_r = 250$ realizations of such noise and apply different imaging algorithms. Figure 2.4 presents the results of computational experiments. It clearly shows that the topological derivative based functional performs as good as Kirchhoff migration and much better than MUSIC and backpropagation, specially at high levels of electronic noise.

We now suppose that the medium bulk modulus is randomly heterogeneous around a constant background: $K^{-1}(\boldsymbol{x}) = 1 + (K^{-1} - 1)\mathbf{1}_D(\boldsymbol{x}) + \mu(\boldsymbol{x})$.

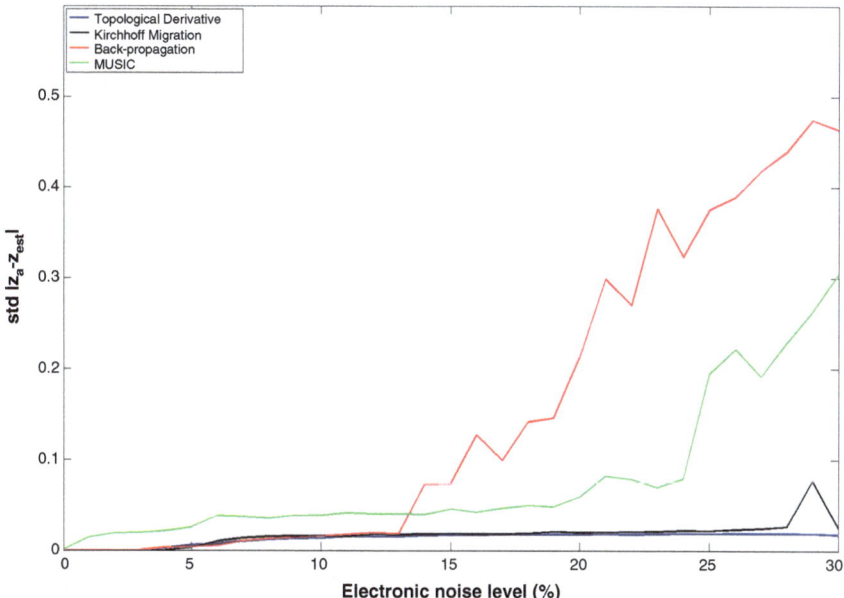

Fig. 2.4. Standard deviations of localization error with respect to electronic noise level for $\mathcal{I}_{\mathrm{MU}}, \mathcal{I}_{\mathrm{BP}}, \mathcal{I}_{\mathrm{KM}},$ and $\mathcal{I}_{\mathrm{TD}}$ with $n = 50$

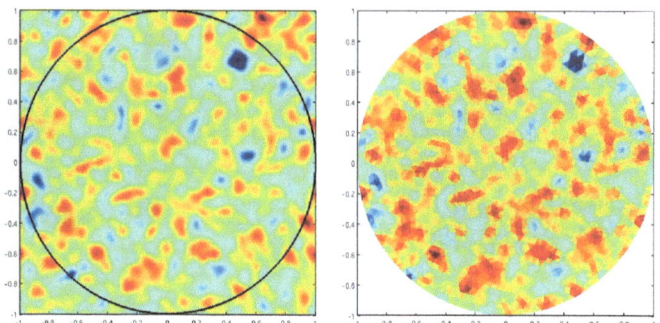

Fig. 2.5. Realization of a medium noise

To simulate μ, we first generate on a grid a white Gaussian noise. Then we filter the Gaussian noise in the Fourier domain by applying a low-pass wavenumber filter. The parameters of the filter are linked to the correlation length l_{μ} of μ [17]. Figure 2.5 shows a typical realization of a medium noise and its projection on the finite-element mesh.

Comparisons between the standard deviations of the localization error with respect to clutter noise for the discussed imaging algorithms are given in Fig. 2.6. Again, the topological derivative based imaging functional is the most robust functional.

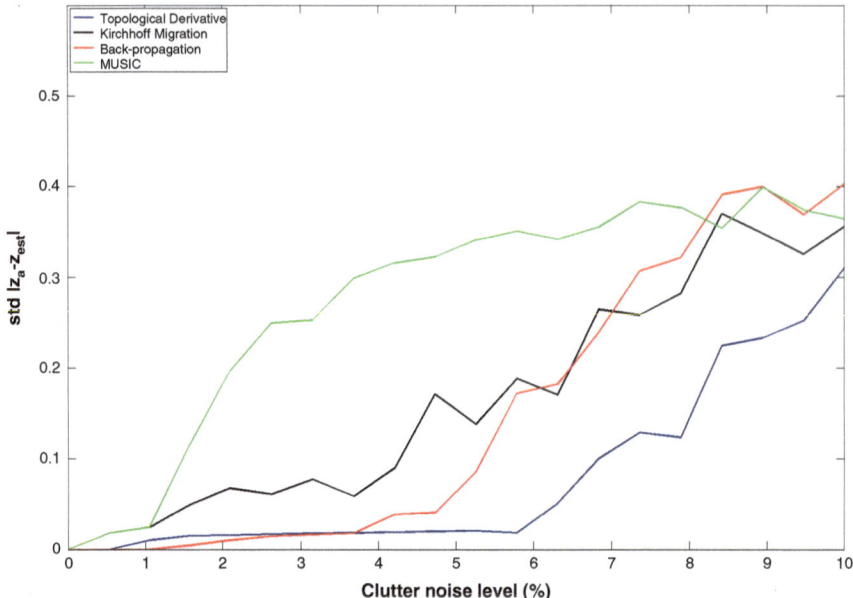

Fig. 2.6. Standard deviations of localization error with respect to clutter noise for $\mathcal{I}_{MU}, \mathcal{I}_{BP}, \mathcal{I}_{KM}$, and $\mathcal{I}_{TD}$ with $n = 50$

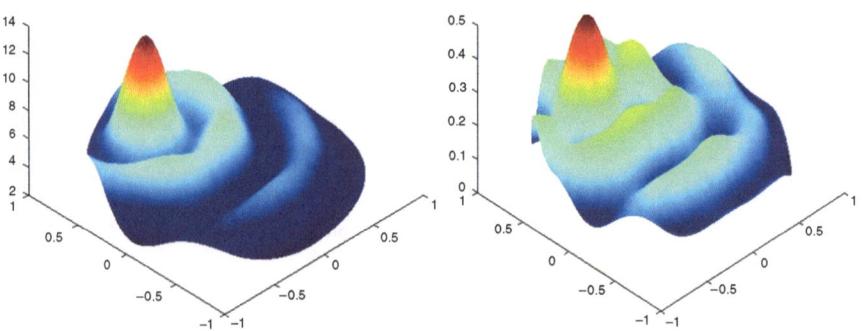

Fig. 2.7. Plots of $\mathcal{I}_{TD}(z^S, \omega)$ with (*left*) $n = 50$ and (*right*) $n = 2$

Density Contrast Only

Here, the parameters of the anomaly are $\rho = 1/2$ and $K = 1$.

Resolution in the Absence of Noise

As shown in Fig. 2.7, the topological derivative based imaging functional $\mathcal{I}_{TD}(z^S, \omega)$ reaches its maximum at the location of the anomaly.

Figure 2.8 shows MUSIC images. As expected from the structure of the response matrix with $K = 1$ (ρ contrast only), its SVD yields two significant singular values [1, 6, 15]. Thus, the illumination vector $\boldsymbol{g}^{(3)}$ does not belong to the signal subspace of the response matrix $\mathbf{A}$. As before, using this vector in the projection step generates a blurred MUSIC peak (figure on the left). To get a sharp peak, we should only project the illumination vectors $\boldsymbol{g}^{(1)}$ and $\boldsymbol{g}^{(2)}$ (figure on the right).

As shown in Fig. 2.9, the backpropagation image has the expected behavior and reaches its maximum at the location of the anomaly.

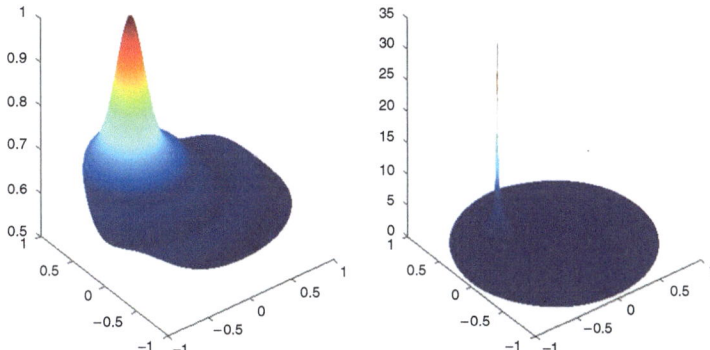

Fig. 2.8. *Left*: MUSIC image using projection of $\boldsymbol{g}^{(1)}, \boldsymbol{g}^{(2)}, \boldsymbol{g}^{(3)}$ on the signal subspace of $\mathbf{A}$. *Right*: image using projection of $\boldsymbol{g}^{(1)}$ and $\boldsymbol{g}^{(2)}$ on the signal subspace of $\mathbf{A}$

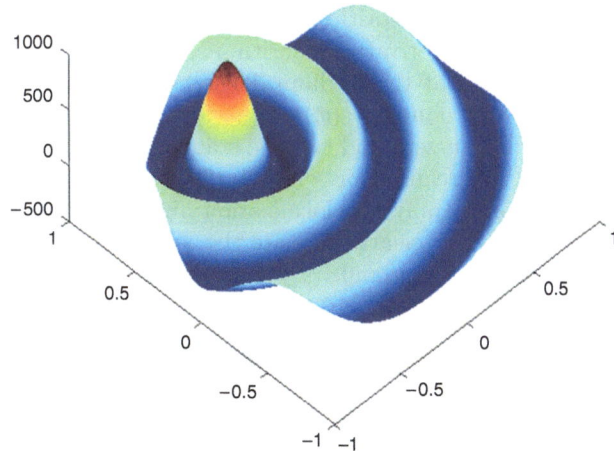

Fig. 2.9. Backpropagation image

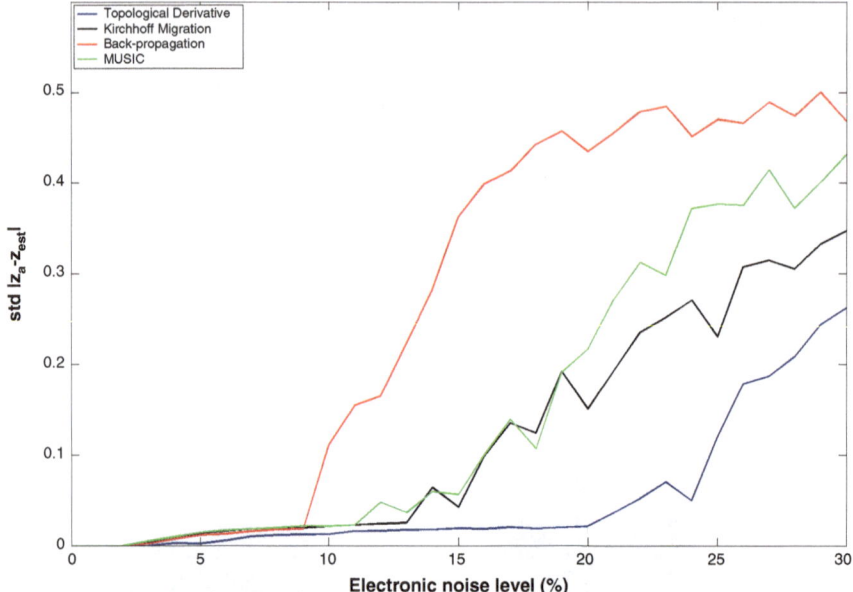

Fig. 2.10. Standard deviations of localization error with respect to electronic noise level for $\mathcal{I}_{\mathrm{MU}}, \mathcal{I}_{\mathrm{BP}}, \mathcal{I}_{\mathrm{KM}}$, and $\mathcal{I}_{\mathrm{TD}}$ with $n = 50$

Stability with Respect to Measurement and Medium Noises

We carry out the same analysis as in the case of only a bulk modulus contrast. Figure 2.10 gives the standard deviation of the localization error as function of the noise level σ for each algorithm.

Again, the topological derivative algorithm seems to be the most robust.

Finally, we suppose that the medium density is randomly heterogeneous around a constant background: $\rho^{-1}(\boldsymbol{x}) = 1 + (\rho^{-1} - 1)\mathbf{1}_D(\boldsymbol{x}) + \mu(\boldsymbol{x})$, with μ a random process of mean zero and tunable standard deviation σ. As before, we compute $N_r = 250$ realizations of such clutter and the corresponding measurements. We then apply the localization algorithms. Stability results are given in Fig. 2.11. They clearly indicate the robustness of the topological derivative based imaging functional.

2.4.2 Measurements at Multiple Frequencies

We illustrate the multifrequency approach with a bulk modulus contrast inclusion. We choose $m = 30$ frequencies between 0 and $\omega_M = 6$ and keep $n = 50$ equidistributed illuminations.

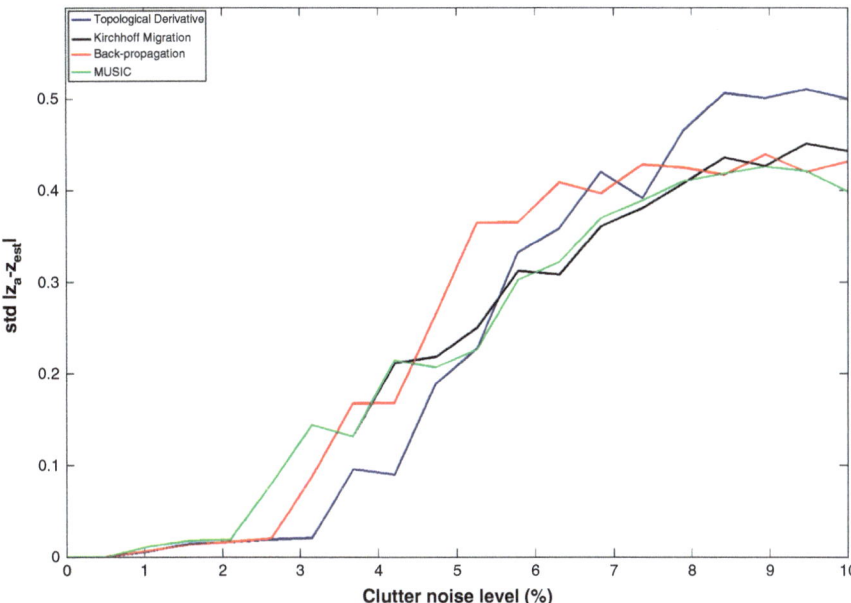

Fig. 2.11. Standard deviations of the localization error with respect to clutter noise for $\mathcal{I}_{\mathrm{MU}}, \mathcal{I}_{\mathrm{BP}}, \mathcal{I}_{\mathrm{KM}}$, and $\mathcal{I}_{\mathrm{TD}}$ with $n = 50$

Measurement Noise

We solve the Helmholtz equation for the n illuminations and m frequencies. We then add a white Gaussian noise with standard deviation σ ranging from 0 to 120% of the L^2 norm of the measurements. We apply the algorithms (Kirchhoff Migration MF, Matched Field MF and Topological Derivative TDF) and plot as previously the standard deviation of the localization error $|z_{\mathrm{est}} - z_a|$ as of function of the noise level for $N_r = 250$ realizations.

Figure 2.12 shows the robustness of $\mathcal{I}_{\mathrm{TDF}}$. We also see from Fig. 2.12 that the multifrequency approach yields an improvement by a factor of $\sqrt{m}$ in the SNR. Here the noise acts incoherently on the amplitude of the signal but not on its phase. Averaging multifrequency measurements cancels such incoherent noise. Moreover, given the expression (2.20) of the SNR of Kirchhoff Migration as a function of ω, it is expected that we should observe even slightly better stability if we repeat m experiments at the highest available frequency available and use $\mathcal{I}_{\mathrm{KMF}}$ for image reconstruction.

Medium Noise

We simulate a clutter as described previously with σ still ranging from 0 to 10% of the mean value of the process μ. Here, measurements at each frequency

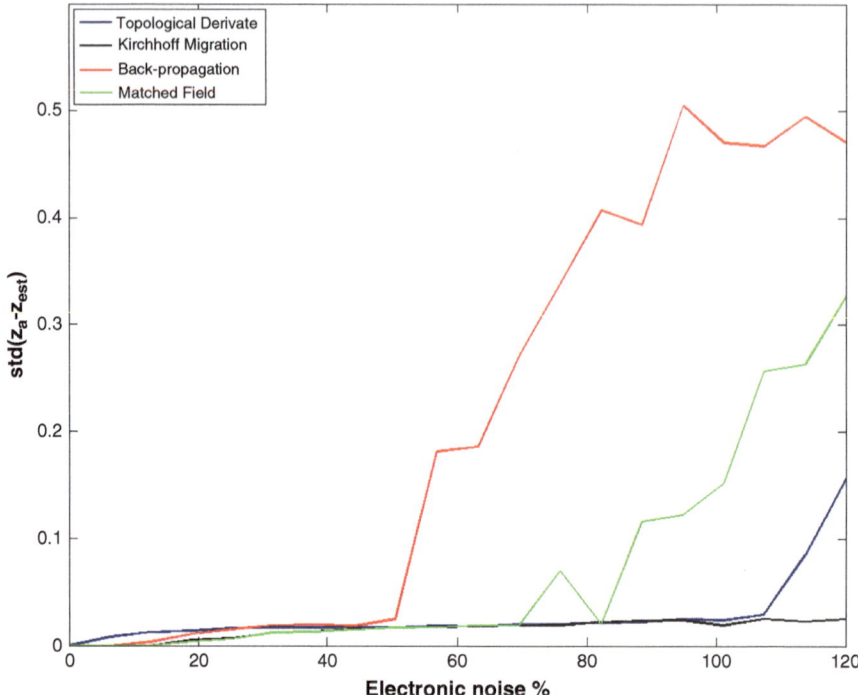

Fig. 2.12. Standard deviations of the localization error with respect to measurement noise for $\mathcal{I}_{BPF}, \mathcal{I}_{KMF}, \mathcal{I}_{MF}$, and $\mathcal{I}_{TDF}$ with $m = 30$ and $n = 50$

will be obtained in the same noisy medium. We cannot observe noise cancellation because the noise induced by such clutter is coherent and acts on the phase of the signal. Indeed, taking multifrequency measurements and bluntly summing is not what should be done. Given the SNR of $\mathcal{I}_{TD}$ in this case [3], we can compute the frequency for which we will get the most stable image. It is the frequency at which the SNR is maximal:

$$\omega_{best} = \frac{2}{\sigma}\sqrt{\frac{2}{l_\mu L_\mu}},$$

where l_μ is the correlation length of μ and L_μ is the propagation distance through the clutter. Since in our case l_μ and σ are small and L_μ is of order one, ω_{best} is out of the chosen frequency range. Our best shot is then to use $\omega = \omega_{max}$.

Figure 2.13 illustrates the previous observation: summing over frequencies the topological derivative imaging functional seems to be counter-productive for a clutter noise medium. A solution to make use of multifrequency measurements in the case of noisy medium could be the CINT inspired approach described previously. However, the computations become really

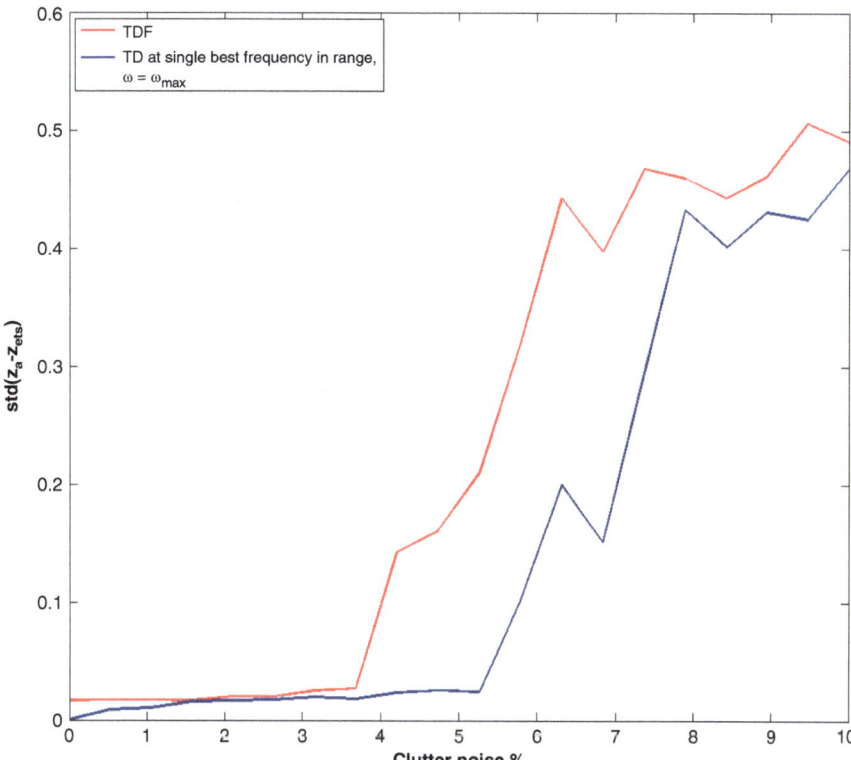

Fig. 2.13. Standard deviations of the localization error with respect to medium noise for $\mathcal{I}_{\mathrm{TDF}}$ with $m = 30$ and $n = 50$ and $\mathcal{I}_{\mathrm{TD}}$ at $\omega_{\max}$ for $n = 50$

heavy and tuning of the parameters Ω_D, X_D really fine, preventing a large scale stability analysis as the one we carried out for other direct algorithms.

2.5 Conclusion

In this chapter we have compared the performance of direct anomaly detection functionals at a fixed as well as multiple frequencies. We have carried out a numerical stability and resolution analysis in the presence of medium and measurement noises. In the case of multifrequency measurements, we have proposed a CINT-type imaging functional to correct the effect of an unknown cluttered bulk modulus (random fluctuations around a known constant) on anomaly detection. We have proved that the topological derivative imaging functional is quite robust with respect to both measurement and medium noises.

Acknowledgments

This work was supported by the ERC Advanced Grant Project MULTIMOD–267184.

References

1. H. Ammari, An Introduction to Mathematics of Emerging Biomedical Imaging, Mathematics & Applications, Vol. 62, Springer, Berlin (2008)
2. H. Ammari, E. Bretin, J. Garnier, V. Jugnon, Coherent interferometry algorithms for photoacoustic imaging, SIAM J. Numer. Anal., to appear.
3. H. Ammari, J. Garnier, V. Jugnon, H. Kang, Stability and resolution analysis for a topological derivative based imaging functional, SIAM J. Control Opt., to appear.
4. H. Ammari, J. Garnier, H. Kang, W.-K. Park, K. Sølna, Imaging schemes for perfectly conducting cracks, SIAM J. Appl. Math. **71**, 68–91 (2011)
5. H. Ammari, J. Garnier, K. Sølna, A statistical approach to target detection and localization in the presence of noise, Waves in Random and Complex Media, to appear.
6. H. Ammari, E. Iakovleva, D. Lesselier, A MUSIC algorithm for locating small inclusions buried in a half-space from the scattering amplitude at a fixed frequency, Multiscale Model. Simul. **3**, 597–628 (2005)
7. H. Ammari, H. Kang, Reconstruction of Small Inhomogeneities from Boundary Measurements, Lecture Notes in Mathematics, Vol. 1846, Springer, Berlin (2004)
8. H. Ammari, H. Kang, Polarization and Moment Tensors: with Applications to Inverse Problems and Effective Medium Theory, Applied Mathematical Sciences, Vol. 162, Springer, New York (2007)
9. H. Ammari, H. Kang, Expansion Methods, *Handbook of Mathematical Methods in Imaging*, 447–499, Springer, New York (2011)
10. H. Ammari, H. Kang, H. Lee, Layer Potential Techniques in Spectral Analysis, Mathematical Surveys and Monographs, Vol. 153, American Mathematical Society, Providence RI (2009)
11. N. Bleistein, J.K. Cohen, J.W. Stockwell Jr., *Mathematics of Multidimensional Seismic Imaging, Migration, and Inversion*, Springer, New York (2001)
12. M. Bonnet, B.B. Guzina, Sounding of finite solid bodies by way of topological derivative, Int. J. Numer. Meth. Eng. **61**, 2344–2373 (2004)
13. L. Borcea, G. Papanicolaou, C. Tsogka, Theory and applications of time reversal and interferometric imaging, Inverse Probl. **19**, 134–164 (2003)
14. L. Borcea, G. Papanicolaou, C. Tsogka, Interferometric array imaging in clutter, Inverse Prob. **21**, 1419–1460 (2005)
15. M. Brühl, M. Hanke, M.S. Vogelius, A direct impedance tomography algorithm for locating small inhomogeneities, Numer. Math. **93**, 635–654 (2003)
16. J. Garnier, Use of random matrix theory for target detection, localization, and reconstruction, Contemporary Math., **548**, 1–20, Amer. Math. Soc., (2011)
17. L. Klimes, Correlation functions of random media, Pure Appl. Geophys. **159**, 1811–1831 (2002)
18. G.W. Milton, The Theory of Composites, Cambridge Monographs on Applied and Computational Mathematics, Cambridge University Press (2001)

19. M.S. Vogelius, D. Volkov, Asymptotic formulas for perturbations in the electro-magnetic fields due to the presence of inhomogeneities, Math. Model. Numer. Anal. **34**, 723–748 (2000)
20. D. Volkov, Numerical methods for locating small dielectric inhomogeneities, Wave Motion **38**, 189–206 (2003)

3

Photoacoustic Imaging for Attenuating Acoustic Media

Habib Ammari[*,1], Elie Bretin[2], Vincent Jugnon[1], and Abdul Wahab[2]

[1] Department of Mathematics and Applications, Ecole Normale Supérieure,
45 Rue d'Ulm, 75005 Paris, France habib.ammari@ens.fr,
vincent.jugnon@ens.fr
[2] Centre de Mathématiques Appliquées, CNRS UMR 7641, Ecole Polytechnique,
91128 Palaiseau, France bretin@cmap.polytechnique.fr,
wahab@cmap.polytechnique.fr

Summary. The aim of this chapter is to consider two challenging problems in photo-acoustic imaging. We consider extended optical sources in an attenuating acoustic background. We provide algorithms to correct the effects of imposed boundary conditions and that of attenuation as well. By testing our measurements against an appropriate family of functions, we show that we can access the Radon transform of the initial condition in the acoustic wave equation, and thus recover quantitatively the absorbing energy density. We also show how to compensate the effect of acoustic attenuation on image quality by using the stationary phase theorem.

3.1 Introduction

In photo-acoustic imaging, optical energy absorption causes thermo-elastic expansion of the tissue, which leads to the propagation of a pressure wave. This signal is measured by transducers distributed on the boundary of the object, which in turn is used for imaging optical properties of the object. The major contribution of photo-acoustic imaging is to provide images of optical contrasts (based on the optical absorption) with the resolution of ultrasound [32].

If the medium is acoustically homogeneous and has the same acoustic properties as the free space, then the boundary of the object plays no role and the optical properties of the medium can be extracted from measurements of the pressure wave by inverting a spherical or a circular Radon transform [16, 17, 21, 22].

In some settings, free space assumptions does not hold. For example, in brain imaging, the skull plays an important acoustic role, and in small animal imaging devices, the metallic chamber may have a strong acoustic effect. In those cases, one has to account for boundary conditions. If a boundary

H. Ammari (ed.), *Mathematical Modeling in Biomedical Imaging II*,
Lecture Notes in Mathematics 2035, DOI 10.1007/978-3-642-22990-9_3,
© Springer-Verlag Berlin Heidelberg 2012

condition has to be imposed on the pressure field, then an explicit inversion formula no longer exists. However, using a quite simple duality approach, one can still reconstruct the optical absorption coefficient. In fact, in the recent works [2, 3], we have investigated quantitative photoacoustic imaging in the case of a bounded medium with imposed boundary conditions. In a further study [1], we proposed a geometric-control approach to deal with the case of limited view measurements. In both cases, we focused on a situation with small optical absorbers in a non-absorbing background and proposed adapted algorithms to locate the absorbers and estimate their absorbed energy.

A second challenging problem in photo-acoustic imaging is to take into account the issue of modelling the acoustic attenuation and its compensation. This subject is addressed in [7,19,20,24,25,27,28,30]. See Chap. 4 for a detailed discussion on the attenuation models and their causality properties.

In this chapter, we propose a new approach to image extended optical sources from photo-acoustic data and to correct the effect of acoustic attenuation. By testing our measurements against an appropriate family of functions, we show that we can access the Radon transform of the initial condition, and thus recover quantitatively any initial condition for the photoacoustic problem. We also show how to compensate the effect of acoustic attenuation on image quality by using the stationary phase theorem. We use a frequency power-law model for the attenuation losses.

The chapter is organized as follows. In Sect. 3.2 we consider the photo-acoustic imaging problem in free space. We first propose three algorithms to recover the absorbing energy density from limited-view and compare their speeds of convergence. We then present two approaches to correct the effect of acoustic attenuation. We use a power-law model for the attenuation. We test the singular value decomposition approach proposed in [24] and provide a new a technique based on the stationary phase theorem. The stationary phase theorem allows us to compute (approximately in terms of the attenuation coefficient) the unattenuated wave from the attenuated one. The ill-posedness character of such an inverse problem will be investigated in detail in Chap. 4. Section 3.3 is devoted to correct the effect of imposed boundary conditions. By testing our measurements against an appropriate family of functions, we show how to obtain the Radon transform of the initial condition in the acoustic wave equation, and thus recover quantitatively the absorbing energy density. We also show how to compensate the effect of acoustic attenuation on image quality by using again the stationary phase theorem. The chapter ends with a discussion.

3.2 Photo-Acoustic Imaging in Free Space

In this section, we first formulate the imaging problem in free space and present a simulation for the reconstruction of the absorbing energy density using the spherical or circular Radon transform. Then, we provide a total variation regularization to find a satisfactory solution of the imaging problem with limited-view data. Finally, we present algorithms for compensating the effect of acoustic attenuation. The main idea is to express the effect of attenuation as a convolution operator. Attenuation correction is then achieved by inverting this operator. Two strategies are used for such deconvolution. The first one is based on the singular value decomposition of the operator and the second one uses its asymptotic expansion based on the stationary phase theorem. We compare the performances of the two approaches.

3.2.1 Mathematical Formulation

We consider the wave equation in $\mathbb{R}^d$,

$$\frac{1}{c_0^2}\frac{\partial^2 p}{\partial t^2}(x,t) - \Delta p(x,t) = 0 \quad \text{in } \mathbb{R}^d \times (0,T),$$

with

$$p(x,0) = p_0 \quad \text{and} \quad \frac{\partial p}{\partial t}(x,0) = 0.$$

Here c_0 is the phase velocity in a non-attenuating medium.

Assume that the support of p_0, the absorbing energy density, is contained in a bounded set Ω of $\mathbb{R}^d$. Our objective in this part is to reconstruct p_0 from the measurements $g(y,t) = p(y,t)$ on $\partial\Omega \times (0,T)$, where $\partial\Omega$ denotes the boundary of Ω.

The problem of reconstructing p_0 is related to the inversion of the spherical Radon transform given by

$$\mathcal{R}_\Omega[f](y,r) = \int_S rf(y + r\xi)\, d\sigma(\xi), \quad (y,r) \in \partial\Omega \times \mathbb{R}^+,$$

where S denotes the unit sphere in $\mathbb{R}^d$. It is known that in dimension $d = 2$, Kirchhoff's formula implies that [14]

$$\begin{cases} p(y,t) = \dfrac{1}{2\pi}\partial_t \displaystyle\int_0^t \dfrac{\mathcal{R}_\Omega[p_0](y,c_0 r)}{\sqrt{t^2 - r^2}}\, dr, \\[2ex] \mathcal{R}_\Omega[p_0](y,r) = 4r \displaystyle\int_0^r \dfrac{p(y,t/c_0)}{\sqrt{r^2 - t^2}}\, dt. \end{cases}$$

Let the operator $\mathcal{W}$ be defined by

$$\mathcal{W}[g](y,r) = 4r\int_0^r \frac{g(y,t/c_0)}{\sqrt{r^2 - t^2}}\, dt \quad \text{for all } g : \partial\Omega \times \mathbb{R}^+ \to \mathbb{R}. \tag{3.1}$$

Then, it follows that

$$\mathcal{R}_\Omega[p_0](y,r) = \mathcal{W}[p](y,r). \tag{3.2}$$

In recent works, a large class of inversion retroprojection formulae for the spherical Radon transform have been obtained in even and odd dimensions when Ω is a ball, see for instance [13, 14, 23, 26]. In dimension 2 when Ω is the unit disk, it turns out that

$$p_0(x) = \frac{1}{(4\pi^2)} \int_{\partial\Omega} \int_0^2 \left[\frac{d^2}{dr^2} \mathcal{R}_\Omega[p_0](y,r) \right] \ln|r^2 - (y-x)^2| \, dr \, d\sigma(y). \quad (3.3)$$

This formula can be rewritten as follows:

$$p_0(x) = \frac{1}{4\pi^2} \mathcal{R}_\Omega^* \mathcal{B} \mathcal{R}_\Omega[p_0](x), \quad (3.4)$$

where $\mathcal{R}_\Omega^*$ is the adjoint of $\mathcal{R}_\Omega$,

$$\mathcal{R}_\Omega^*[g](x) = \int_{\partial\Omega} g(y, |y-x|) \, d\sigma(y),$$

and $\mathcal{B}$ is defined by

$$\mathcal{B}[g](x,t) = \int_0^2 \frac{d^2 g}{dr^2}(y,r) \ln(|r^2 - t^2|) \, dr$$

for $g : \Omega \times \mathbb{R}^+ \to \mathbb{R}$.

In Fig. 3.1, we give a numerical illustration for the reconstruction of p_0 using the spherical Radon transform. We adopt the same approach as in [13] for the discretization of formulae (3.1) and (3.3). Note that in the numerical examples presented in this section, N_θ denotes the number of equally spaced angles on $\partial\Omega$, the pressure signals are uniformly sampled at N time steps, and the phantom (the initial pressure distribution p_0) is sampled on a uniform Cartesian grid with $N_R \times N_R$ points.

3.2.2 Limited-View Data

In many situations, we have only at our disposal data on $\Gamma \times (0, T)$, where $\Gamma \subset \partial\Omega$. As illustrated in Fig. 3.2, restricting the integration in formula (3.3) to Γ as follows:

$$p_0(x) \simeq \frac{1}{(4\pi^2)} \int_\Gamma \int_0^2 \left[\frac{d^2}{dr^2} \mathcal{R}_\Omega[p_0](y,r) \right] \ln|r^2 - (y-x)^2| \, dr \, d\sigma(y), \quad (3.5)$$

is not stable enough to give a correct reconstruction of p_0.

The inverse problem becomes severely ill-posed and needs to be regularized (see for instance [15, 33]). We apply here a Tikhonov regularization with a total variation term, which is well adapted to the reconstruction of smooth

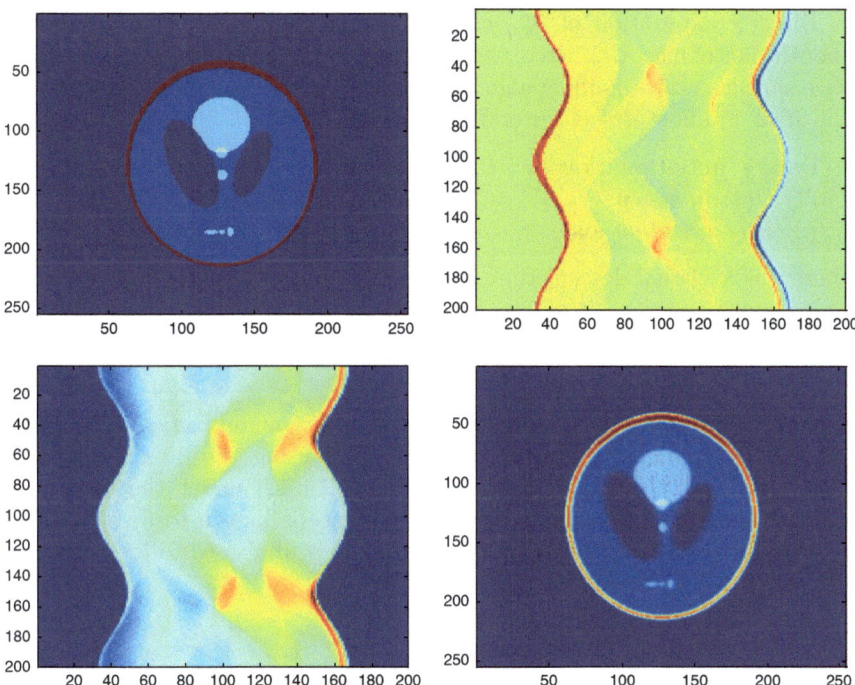

Fig. 3.1. Numerical inversion using (3.3) with $N = 256$, $N_R = 200$ and $N_\theta = 200$. *Top left:* p_0; *Top right:* $p(y,t)$ with $(y,t) \in \partial\Omega \times (0,2)$; *Bottom left:* $\mathcal{R}_\Omega[p_0](y,t)$ with $(y,t) \in \partial\Omega \times (0,2)$; *Bottom right:* $\frac{1}{4\pi^2}\mathcal{R}_\Omega^*\mathcal{B}\mathcal{R}_\Omega[p_0]$

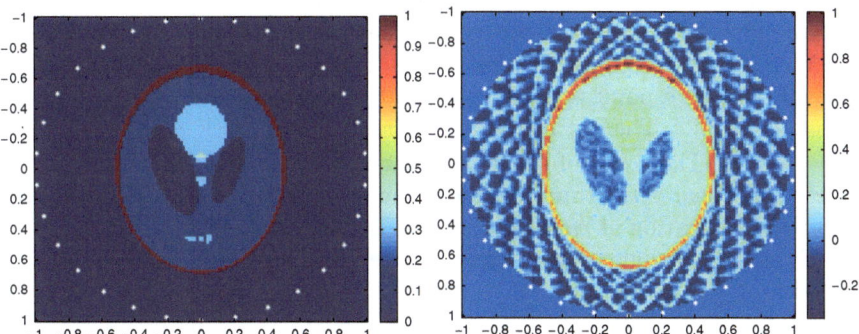

Fig. 3.2. Numerical inversion with truncated (3.3) formula with $N = 128$, $N_R = 128$, and $N_\theta = 30$. *Left:* p_0; *Right:* $\frac{1}{4\pi^2}\mathcal{R}_\Omega^*\mathcal{B}\mathcal{R}_\Omega[p_0]$

solutions with front discontinuities. We then introduce the function $p_{0,\eta}$ as the minimizer of

$$J[f] = \frac{1}{2}\|Q\left[\mathcal{R}_\Omega[f] - g\right]\|^2_{L^2(\partial\Omega\times(0,2))} + \eta\|\nabla f\|_{L^1(\Omega)},$$

where Q is a positive weight operator.

Direct computation of $p_{0,\eta}$ can be complicated as the TV term is not smooth (not of class $\mathcal{C}^1$). Here, we obtain an approximation of $p_{0,\eta}$ via an iterative shrinkage-thresholding algorithm [10, 12]. This algorithm can be viewed as a split, gradient-descent, iterative scheme:

- Data g, initial solution $f_0 = 0$;
- (1) Data link step: $f_{k+1/2} = f_k - \gamma \mathcal{R}_\Omega^* Q^* Q \left[\mathcal{R}_\Omega[f_k] - g \right]$;
- (2) Regularization step: $f_k = T_{\gamma\eta}[f_{k+1/2}]$,

where γ is a virtual descent time step and the operator T_η is given by

$$T_\eta[y] = \arg\min_x \left\{ \frac{1}{2}\|y - x\|_{L^2}^2 + \eta\|\nabla x\|_{L^1} \right\}.$$

Note that T_η defines a proximal point method. One advantage of the algorithm is to minimize implicitly the TV term using the duality algorithm of Chambolle [8]. This algorithm converges [10, 12] under the assumption $\gamma\|\mathcal{R}_\Omega^* Q^* Q \mathcal{R}_\Omega\| \leq 1$, but its rate of convergence is known to be slow. Thus, in order to accelerate the convergence rate, we will also consider the variant algorithm of Beck and Teboulle [6] defined as:

- Data g, initial set: $f_0 = x_0 = 0$, $t_1 = 1$.
- (1) $x_k = T_{\gamma\eta}\left(f_k - \gamma \mathcal{R}_\Omega^* Q^* Q \left[\mathcal{R}_\Omega[f_k] - g \right] \right)$.
- (2) $f_{k+1} = x_k + \frac{t_k - 1}{t_{k+1}}(x_k - x_{k-1})$ with $t_{k+1} + \frac{1 + \sqrt{1 + 4t_k^2}}{2}$.

The standard choice of Q is the identity, Id, and then it is easy to see that $\|\mathcal{R}_\Omega \mathcal{R}_\Omega^*\| \simeq 2\pi$. It will also be interesting to use $Q = \frac{1}{2\pi}\mathcal{B}^{1/2}$, which is well defined since $\mathcal{B}$ is symmetric and positive. In this case, $\mathcal{R}_\Omega^* Q^* Q \simeq \mathcal{R}_\Omega^{-1}$ and we can hope to improve the convergence rate of the regularized algorithm.

We compare three algorithms of this kind in Fig. 3.3. The first and the second one correspond to the simplest algorithm with respectively $Q = \mathrm{Id}$ and $Q = \frac{1}{2\pi}\mathcal{B}^{1/2}$. The last method uses the variant of Beck and Teboulle with $Q = \frac{1}{2\pi}\mathcal{B}^{1/2}$. The speed of convergence of each of these algorithms is plotted in Fig. 3.3. Clearly, the third method is the best and after 30 iterations, a very good approximation of p_0 is reconstructed.

Two limited-angle experiments are presented in Fig. 3.4 using the third algorithm.

3.2.3 Compensation of the Effect of Acoustic Attenuation

Our aim in this section is to compensate for the effect of acoustic attenuation. We refer to Chap. 4 for a study of the ill-posedness character of such an inverse problem. The pressures $p(x, t)$ and $p_a(x, t)$ are respectively solutions of the following wave equations:

$$\frac{1}{c_0^2}\frac{\partial^2 p}{\partial t^2}(x, t) - \Delta p(x, t) = \frac{1}{c_0^2}\delta'_{t=0}p_0(x),$$

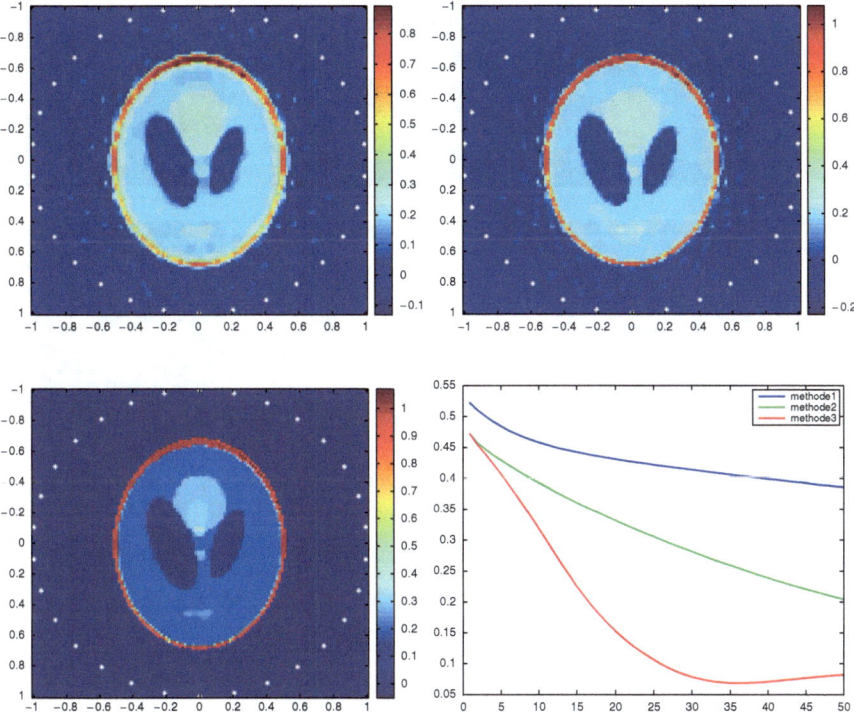

Fig. 3.3. Iterative shrinkage-thresholding solution after 30 iterations with $\eta = 0.01$, $N = 128$, $N_R = 128$, and $N_\theta = 30$. *Top left*: simplest algorithm with $Q = \mathrm{Id}$ and $\mu = 1/(2\pi)$; *Top right*: simplest algorithm with $Q = \frac{1}{2\pi}\mathcal{B}^{1/2}$ and $\mu = 0.5$; *Bottom left*: Beck and Teboulle variant with $Q = \frac{1}{2\pi}\mathcal{B}^{1/2}$ and $\mu = 0.5$; *Bottom right*: error $k \rightarrow \|f_k - p_0\|_\infty$ for each of the previous situations

and

$$\frac{1}{c_0^2}\frac{\partial^2 p_a}{\partial t^2}(x,t) - \Delta p_a(x,t) - L(t) * p_a(x,t) = \frac{1}{c_0^2}\delta'_{t=0}p_0(x),$$

where L is defined by

$$L(t) = \frac{1}{\sqrt{2\pi}}\int_{\mathbb{R}}\left(K^2(\omega) - \frac{\omega^2}{c_0^2}\right)e^{i\omega t}d\omega. \tag{3.6}$$

Many models exist for $K(\omega)$. Here we use the power-law model. Then $K(\omega)$ is the complex wave number, defined by

$$K(\omega) = \frac{\omega}{c(\omega)} + ia|\omega|^\varsigma, \tag{3.7}$$

where ω is the frequency, $c(\omega)$ is the frequency dependent phase velocity and $1 \leq \varsigma \leq 2$ is the power of the attenuation coefficient. See [29]. A common

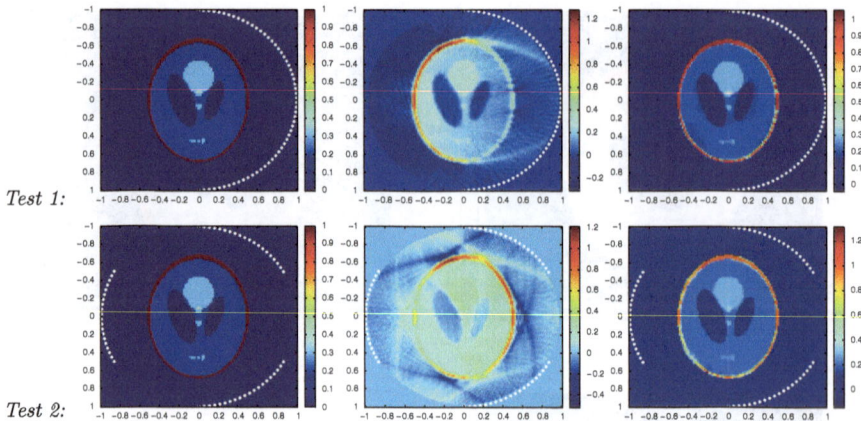

Test 1:

Test 2:

Fig. 3.4. Case of limited angle with Beck and Teboulle iterative shrinkage-thresholding after 50 iterations, with parameters equal to $\eta = 0.01$, $N = 128$, $N_R = 128$, $N_\theta = 64$ and $Q = \frac{1}{2\pi}\mathcal{B}^{1/2}$. *Left:* p_0; *Center:* $\frac{1}{4\pi^2}\mathcal{R}_\Omega^*\mathcal{B}\mathcal{R}_\Omega[p_0]$; *Right:* f_{50}

model, known as the thermo-viscous model, is given by $K(\omega) = \frac{\omega}{c_0\sqrt{1-ia\omega c_0}}$ and corresponds approximately to $\zeta = 2$ with $c(\omega) = c_0$.

Our strategy is now to:

- Estimate $p(y,t)$ from $p_a(y,t)$ for all $(y,t) \in \partial\Omega \times \mathbb{R}^+$.
- Apply the inverse formula for the spherical Radon transform to reconstruct p_0 from the non-attenuated data.

A natural definition of an attenuated spherical Radon transform $\mathcal{R}_{a,\Omega}$ is

$$\mathcal{R}_{a,\Omega}[p_0] = \mathcal{W}[p_a]. \tag{3.8}$$

3.2.4 Relationship Between p and p_a

Recall that the Fourier transforms of p and p_a satisfy

$$\left(\Delta + \left(\frac{\omega}{c_0}\right)^2\right)\hat{p}(x,\omega) = \frac{i\omega}{\sqrt{2\pi}c_0^2}p_0(x) \quad \text{and} \quad \left(\Delta + K(\omega)^2\right)\hat{p}_a(x,\omega)$$

$$= \frac{i\omega}{\sqrt{2\pi}c_0^2}p_0(x),$$

which implies that

$$\hat{p}(x, c_0 K(\omega)) = \frac{c_0 K(\omega)}{\omega}\hat{p}_a(x,\omega).$$

The issue is to estimate p from p_a using the relationship $p_a = \mathcal{L}[p]$, where $\mathcal{L}$ is defined by

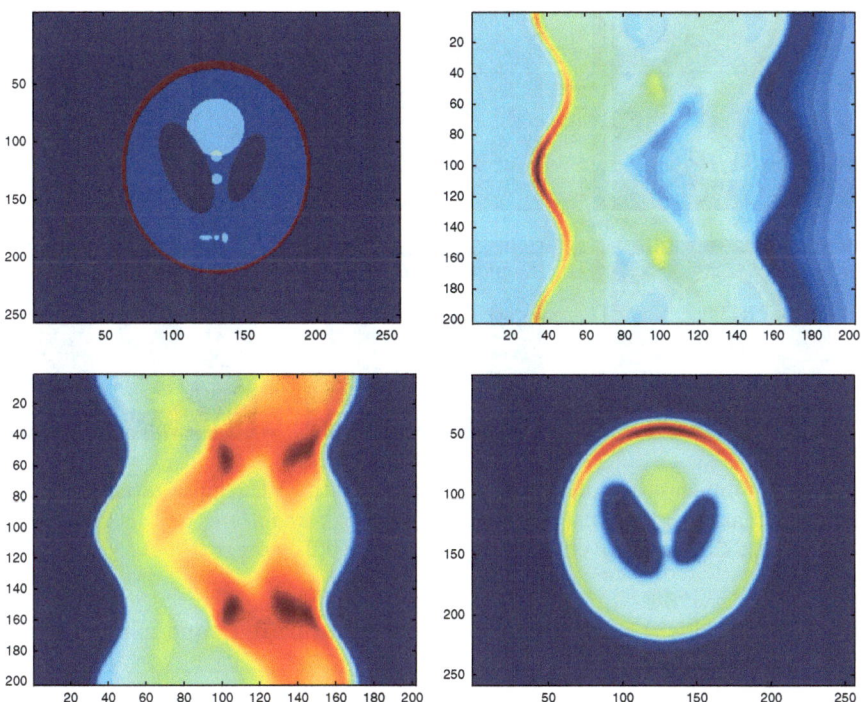

Fig. 3.5. Numerical inversion of attenuated wave equation with $K(\omega) = \frac{\omega}{c_0} + ia\omega^2/2$ and $a = 0.001$. Here $N = 256$, $N_R = 200$ and $N_\theta = 200$. *Top left:* p_0; *Top right:* $p_a(y,t)$ with $(y,t) \in \partial\Omega \times (0,2)$; *Bottom left:* $\mathcal{W}p_a(y,t)$ with $(y,t) \in \partial\Omega \times (0,2)$; *Bottom right:* $\frac{1}{4\pi^2}\mathcal{R}_\Omega^*\mathcal{B}\left(\mathcal{W}[p_a](y,t)\right)$

$$\mathcal{L}[\phi](s) = \frac{1}{2\pi} \int_{\mathbb{R}} \frac{\omega}{c_0 K(\omega)} e^{-i\omega s} \int_0^\infty \phi(t) e^{ic_0 K(\omega)t} \, dt \, d\omega.$$

The main difficulty is that $\mathcal{L}$ is not well conditioned. We will compare two approaches. The first one uses a regularized inverse of $\mathcal{L}$ via a singular value decomposition (SVD), which has been recently introduced in [24]. The second one is based on the asymptotic behavior of $\mathcal{L}$ as the attenuation coefficient a tends to zero.

Figure 3.5 gives some numerical illustrations of the inversion of the attenuated spherical Radon transform without a correction of the attenuation effect, where a thermo-viscous attenuation model is used with $c_0 = 1$.

3.2.5 A SVD Approach

La Rivière, Zhang, and Anastasio have recently proposed in [24] to use a regularized inverse of the operator $\mathcal{L}$ obtained by a standard SVD approach:

$$\mathcal{L}[\phi] = \sum_l \sigma_l \langle \phi, \tilde{\psi}_l \rangle \psi_l,$$

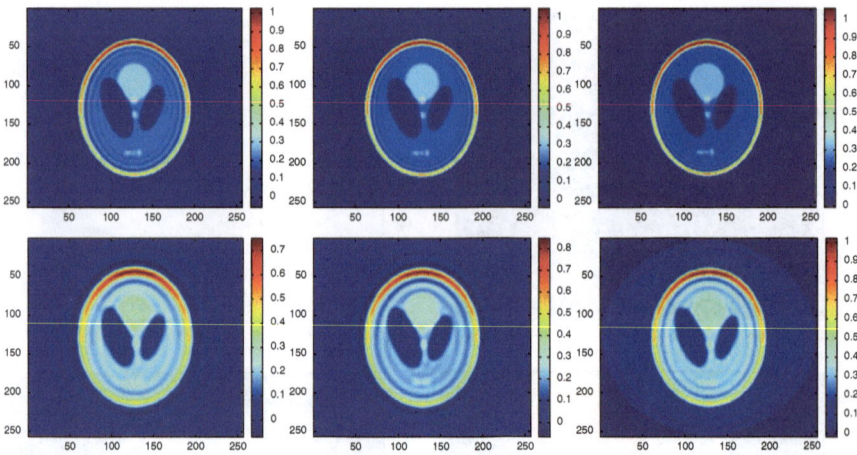

Fig. 3.6. Compensation of acoustic attenuation with SVD regularization: $N = 256$, $N_R = 200$ and $N_\theta = 200$. *First line:* $a = 0.0005$; *second line:* $a = 0.0025$. *Left to right:* using $\mathcal{L}_{1,\epsilon}^{-1}$ respectively with $\epsilon = 0.01$, $\epsilon = 0.001$ and $\epsilon = 0.0001$

where $(\tilde{\psi}_l)$ and (ψ_l) are two orthonormal bases of $L^2(0, T)$ and σ_l are positives eigenvalues such that

$$
\begin{cases}
\mathcal{L}^*[\phi] & = \sum_l \sigma_l \langle \phi, \psi_l \rangle \tilde{\psi}_l, \\
\mathcal{L}^*\mathcal{L}[\phi] & = \sum_l \sigma_l^2 \langle \phi, \tilde{\psi}_l \rangle \tilde{\psi}_l, \\
\mathcal{L}\mathcal{L}^*[\phi] & = \sum_l \sigma_l^2 \langle \phi, \psi_l \rangle \psi_l.
\end{cases}
$$

An ϵ-approximation inverse of $\mathcal{L}$ is then given by

$$
\mathcal{L}_{1,\epsilon}^{-1}[\phi] = \sum_l \frac{\sigma_l}{\sigma_l^2 + \epsilon^2} \langle \phi, \psi_l \rangle \tilde{\psi}_l,
$$

where $\epsilon > 0$.

In Fig. 3.6 we present some numerical inversions of the thermo-viscous wave equation with $a = 0.0005$ and $a = 0.0025$. We first obtain the ideal measurements from the attenuated ones and then apply the inverse formula for the spherical Radon transform to reconstruct p_0 from the ideal data. We take ϵ respectively equal to 0.01, 0.001 and 0.0001. The operator $\mathcal{L}$ is discretized to obtain an $N_R \times N_R$ matrix to which we apply an SVD decomposition. A regularization of the SVD allows us to construct $\mathcal{L}_{1,\epsilon}^{-1}$.

As expected, this algorithm corrects a part of the attenuation effect but is unstable when ϵ tends to zero.

3.2.6 Asymptotics of $\mathcal{L}$

In physical situations, the coefficient of attenuation a is very small. We will take into account this phenomenon and introduce an approximation of $\mathcal{L}$ and

$\mathcal{L}^{-1}$ as a goes to zero:

$$\mathcal{L}_k[\phi] = \mathcal{L}[\phi] + o(a^{k+1}) \quad \text{and} \quad \mathcal{L}_{2,k}^{-1}[\phi] = \mathcal{L}^{-1}[\phi] + o(a^{k+1}),$$

where k represents an order of approximation.

Thermo-Viscous Case: $K(\omega) = \frac{\omega}{c_0} + ia\omega^2/2$

Let us consider in this section the attenuation model $K(\omega) = \frac{\omega}{c_0} + ia\omega^2/2$ at low frequencies $\omega \ll \frac{1}{a}$, such that

$$\frac{1}{1 + iac_0\omega/2} \simeq 1 - i\frac{ac_0}{2}\omega.$$

The operator $\mathcal{L}$ is approximated as follows

$$\mathcal{L}[\phi](s) \simeq \frac{1}{2\pi} \int_0^\infty \phi(t) \int_{\mathbb{R}} \left(1 - i\frac{ac_0}{2}\omega\right) e^{-\frac{1}{2}c_0a\omega^2 t} e^{i\omega(t-s)} \, d\omega \, dt.$$

Since

$$\frac{1}{\sqrt{2\pi}} \int_{\mathbb{R}} e^{-\frac{1}{2}c_0a\omega^2 t} e^{i\omega(t-s)} d\omega = \frac{1}{\sqrt{c_0 at}} e^{-\frac{1}{2}\frac{(s-t)^2}{c_0 at}},$$

and

$$\frac{1}{\sqrt{2\pi}} \int_{\mathbb{R}} \frac{-iac_0\omega}{2} e^{-\frac{1}{2}c_0a\omega^2 t} e^{i\omega(t-s)} d\omega = \frac{ac_0}{2}\partial_s\left(\frac{1}{\sqrt{c_0 at}} e^{-\frac{1}{2}\frac{(s-t)^2}{c_0 at}}\right),$$

it follows that

$$\mathcal{L}[\phi] \simeq \left(1 + \frac{ac_0}{2}\partial_s\right)\left(\frac{1}{\sqrt{2\pi}} \int_0^{+\infty} \phi(t)\frac{1}{\sqrt{c_0 at}} e^{-\frac{1}{2}\frac{(s-t)^2}{c_0 at}} \, dt\right).$$

We then investigate the asymptotic behavior of $\tilde{\mathcal{L}}$ defined by

$$\tilde{\mathcal{L}}[\phi] = \frac{1}{\sqrt{2\pi}} \int_0^{+\infty} \phi(t)\frac{1}{\sqrt{c_0 at}} e^{-\frac{1}{2}\frac{(s-t)^2}{c_0 at}} \, dt. \tag{3.9}$$

Since the phase in (3.9) is quadratic and a is small, by the stationary phase theorem (see Appendix 1) we can prove that

$$\tilde{\mathcal{L}}[\phi](s) = \sum_{i=0}^k \frac{(c_0 a)^i}{2^i i!} D_i[\phi](s) + o(a^k), \tag{3.10}$$

where the differential operators D_i satisfy $D_i[\phi](s) = (t^i\phi(t))^{(2i)}(s)$. See Appendix 2. We can also deduce the following approximation of order k of $\tilde{\mathcal{L}}^{-1}$

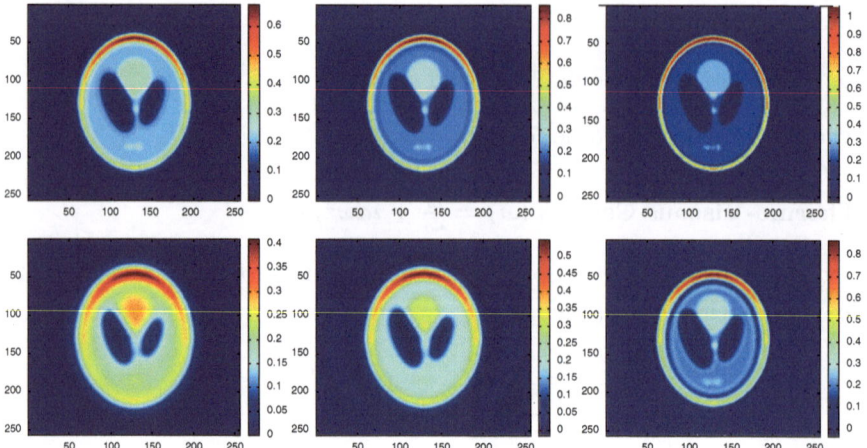

Fig. 3.7. Compensation of acoustic attenuation with formula (3.12): $N = 256$, $N_R = 200$ and $N_\theta = 200$. *First line*: $a = 0.0005$; *second line*: $a = 0.0025$. *Left*: $\tilde{\mathcal{L}}_k^{-1}$ with $k = 0$; *Center*: $\tilde{\mathcal{L}}_k^{-1}$ with $k = 1$; *Right*: $\tilde{\mathcal{L}}_k^{-1}$ with $k = 8$

$$\tilde{\mathcal{L}}_k^{-1}[\psi] = \sum_{j=0}^{k} a^j \psi_{k,j}, \tag{3.11}$$

where $\psi_{k,j}$ are defined recursively by

$$\psi_{k,0} = \psi \quad \text{and} \quad \psi_{k,j} = -\sum_{i=1}^{j} \frac{c_0^i}{2^i i!} D_i[\psi_{k,j-i}], \text{ for all } j \le k.$$

Finally, we define

$$\mathcal{L}_k = \left(1 + \frac{ac_0}{2}\partial_s\right)\tilde{\mathcal{L}}_k \quad \text{and} \quad \mathcal{L}_{2,k}^{-1} = \tilde{\mathcal{L}}_k^{-1}\left(1 + \frac{ac_0}{2}\partial_t\right)^{-1}. \tag{3.12}$$

We plot in Figs. 3.6 and 3.7 some numerical reconstructions of p_0 using a thermo-viscous wave equation with $a = 0.0005$ and $a = 0.0025$. We take the value of k respectively equal to $k = 0$, $k = 1$ and $k = 8$. These reconstructions seem to be as good as those obtained by the SVD regularization approach. Moreover, this new algorithm has better stability properties.

General Case: $K(\omega) = \omega + ia|\omega|^\zeta$ with $1 \le \zeta < 2$

We now consider the attenuation model $K(\omega) = \frac{\omega}{c_0} + ia|\omega|^\zeta$ with $1 \le \zeta < 2$. We first note that this model is not causal but can be changed to a causal one; see [9] and Chap. 4. However, since our main purpose here is to give insights for the compensation of the effect of attenuation on image reconstruction, we work with this quite general model because of its simplicity. As before, the problem can be reduced to the approximation of the operator $\tilde{\mathcal{L}}$ defined by

$$\tilde{\mathcal{L}}[\phi](s) = \int_0^\infty \phi(t) \int_{\mathbb{R}} e^{i\omega(t-s)} e^{-|\omega|^\varsigma c_0 a t} d\omega dt.$$

It is also interesting to see that its adjoint $\tilde{\mathcal{L}}^*$ satisfies

$$\tilde{\mathcal{L}}^*[\phi](s) = \int_0^\infty \phi(t) \int_{\mathbb{R}} e^{i\omega(s-t)} e^{-|\omega|^\varsigma c_0 a s} d\omega dt.$$

Suppose for the moment that $\varsigma = 1$, and working with the adjoint operator $\mathcal{L}^*$, we see that

$$\tilde{\mathcal{L}}^*[\phi](s) = \frac{1}{\pi} \int_0^\infty \frac{c_0 a s}{(c_0 a s)^2 + (s-t)^2} \phi(t) dt.$$

Invoking the dominated convergence theorem, we have

$$\lim_{a\to 0} \tilde{\mathcal{L}}^*[\phi](s) = \lim_{a\to 0} \frac{1}{\pi} \int_{-\frac{1}{ac_0}}^\infty \frac{1}{1+y^2} \phi(s + c_0 a y s) dy$$

$$= \frac{1}{\pi} \int_{-\infty}^\infty \frac{1}{1+y^2} \phi(s) dy = \phi(s).$$

More precisely, introducing the fractional Laplacian $\Delta^{1/2}$ as follows

$$\Delta^{1/2}\phi(s) = \frac{1}{\pi} \text{p.v.} \int_{-\infty}^{+\infty} \frac{\phi(t) - \phi(s)}{(t-s)^2} dt,$$

where p.v. stands for the Cauchy principal value, we get

$$\frac{1}{a}\left(\tilde{\mathcal{L}}^*[\phi](s) - \phi(s)\right) = \frac{1}{a} \int_{-\infty}^\infty \frac{1}{\pi c_0 a s} \frac{1}{1 + \left(\frac{s-t}{c_0 a s}\right)^2} (\phi(t) - \phi(s)) dt$$

$$= \int_{-\infty}^\infty \frac{1}{\pi} \frac{c_0 s}{(c_0 a s)^2 + (s-t)^2} (\phi(t) - \phi(s)) dt$$

$$= \lim_{\epsilon\to 0} \int_{\mathbb{R}\setminus[s-\epsilon, s+\epsilon]} \frac{1}{\pi} \frac{c_0 s}{(c_0 a s)^2 + (s-t)^2} (\phi(t) - \phi(s)) dt$$

$$\to \lim_{\epsilon\to 0} \int_{\mathbb{R}\setminus[s-\epsilon, s+\epsilon]} \frac{1}{\pi} \frac{c_0 s}{(s-t)^2} (\phi(t) - \phi(s)) dt$$

$$= c_0 s \Delta^{1/2}\phi(s),$$

as a tends to zero. We therefore deduce that

$$\tilde{\mathcal{L}}^*[\phi](s) = \phi(s) + c_0 a s \Delta^{1/2}\phi(s) + o(a) \quad \text{and} \quad \tilde{\mathcal{L}}^*[\phi](s)$$
$$= \phi(s) + c_0 a \Delta^{1/2}(s\phi(s)) + o(a).$$

Applying exactly the same argument for $1 < \varsigma < 2$, we obtain that

$$\mathcal{L}[\phi](s) = \phi(s) + C c_0 a \Delta^{\varsigma/2}(s\phi(s)) + o(a),$$

where C is a constant, depending only on ς and $\Delta^{\varsigma/2}$, defined by

$$\Delta^{\varsigma/2}\phi(s) = \frac{1}{\pi} \text{p.v.} \int_{-\infty}^{+\infty} \frac{\phi(t) - \phi(s)}{(t-s)^{1+\varsigma}} dt.$$

3.2.7 Iterative Shrinkage-Thresholding Algorithm with Correction of Attenuation

The previous correction of attenuation is not so efficient for a large attenuation coefficient a. In this case, to improve the reconstruction, we may use again a Tikhonov regularization. Let $\mathcal{R}_{\Omega,a,k}^{-1}$ be an approximate inverse of the attenuated spherical Radon transform $\mathcal{R}_{\Omega,a}$:

$$\mathcal{R}_{\Omega,a,k}^{-1} = \mathcal{R}_{\Omega^{-1}} \mathcal{W} \mathcal{L}_{2,k}^{-1} \mathcal{W}^{-1}.$$

Although its convergence is not clear, we will now consider the following iterative shrinkage-thresholding algorithm:

- Data g, initial set: $f_0 = x_0 = 0$, $t_1 = 1$.
- (1) $x_j = T_{\gamma\eta}\left(f_j - \gamma\mathcal{R}_{\Omega,a,k}^{-1}\left(\mathcal{R}_{\Omega,a}f_j - g\right)\right)$.
- (2) $f_{j+1} = x_j + \frac{t_j-1}{t_{j+1}}(x_j - x_{j-1})$ with $t_{j+1} + \frac{1+\sqrt{1+4t_j^2}}{2}$.

  Figure 3.8 shows the efficiency of this algorithm.

3.3 Photo-Acoustic Imaging with Imposed Boundary Conditions

In this section, we consider the case where a boundary condition has to be imposed on the pressure field. We first formulate the photo-acoustic imaging problem in a bounded domain before reviewing the reconstruction procedures. We refer the reader to [31] where the half-space problem has been considered. We then introduce a new algorithm which reduces the reconstruction problem to the inversion of a Radon transform. This procedure is particularly well-suited for extended absorbers. Finally, we discuss the issue of correcting the attenuation effect and propose an algorithm analogous to the one described in the previous section.

3.3.1 Mathematical Formulation

Let Ω be a bounded domain. We consider the wave equation in the domain Ω:

$$\begin{cases} \dfrac{1}{c_0^2}\dfrac{\partial^2 p}{\partial t^2}(x,t) - \Delta p(x,t) = 0 \text{ in } \Omega \times (0,T), \\ p(x,0) = p_0(x) \hspace{2.5cm} \text{in } \Omega, \\ \dfrac{\partial p}{\partial t}(x,0) = 0 \hspace{2.5cm} \text{in } \Omega, \end{cases} \hspace{1cm} (3.13)$$

with the Dirichlet (resp. the Neumann) imposed boundary conditions:

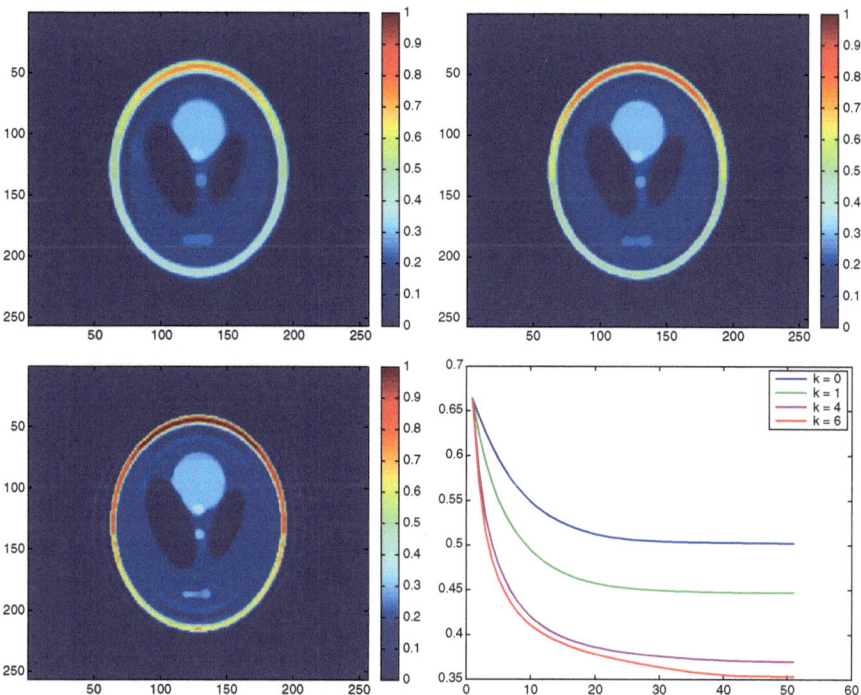

Fig. 3.8. Numerical results using iterative shrinkage-thresholding algorithm with $\eta = 0.001$ and $a = 0.0025$. *Left up:* f_{50} with $k = 0$; *Top right:* f_{50} with $k = 1$; *Bottom left:* f_{50} with $k = 6$; *Bottom right:* error $j \rightarrow \|f_j - p_0\|$ for different values of k

$$p(x,t) = 0 \qquad \left(\text{resp. } \frac{\partial p}{\partial \nu}(x,t) = 0\right) \qquad \text{on } \partial\Omega \times (0,T). \qquad (3.14)$$

Our objective in the next subsection is to reconstruct $p_0(x)$ from the measurements of $\dfrac{\partial p}{\partial \nu}(x,t)$ (resp. $p(x,t)$) on the boundary $\partial\Omega \times (0,T)$.

3.3.2 Inversion Algorithms

Consider probe functions satisfying

$$\begin{cases} \dfrac{1}{c_0^2}\dfrac{\partial^2 v}{\partial t^2}(x,t) - \Delta v(x,t) = 0 & \text{in } \Omega \times (0,T), \\ v(x,T) = 0 & \text{in } \Omega, \\ \dfrac{\partial v}{\partial t}(x,T) = 0 & \text{in } \Omega. \end{cases} \qquad (3.15)$$

Multiplying (3.13) by v and integrating by parts yields (in the case of Dirichlet boundary conditions):

$$\int_0^T \int_{\partial\Omega} \frac{\partial p}{\partial \nu}(x,t)v(x,t)d\sigma(x)dt = \int_\Omega p_0(x)\frac{\partial v}{\partial t}(x,0)dx. \qquad (3.16)$$

Choosing a probe function v with proper initial time derivative allows us to infer information on p_0 (right-hand side in (3.16)) from our boundary measurements (left-hand side in (3.16)).

In [2], considering a full view setting, we used a 2-parameter travelling plane wave given by

$$v_{\tau,\theta}^{(1)}(x,t) = \delta\left(\frac{x\cdot\theta}{c_0}+t-\tau\right), \qquad (3.17)$$

and we determined the inclusions' characteristic functions by varying (θ,τ). We also used in three dimensions the spherical waves given by

$$w_{\tau,y}(x,t) = \frac{\delta\left(t+\tau-\frac{|x-y|}{c_0}\right)}{4\pi|x-y|}, \qquad (3.18)$$

for $y\in\mathbb{R}^3\setminus\Omega$, to probe the medium.

In [1], we assumed that measurements are only made on a part of the boundary $\Gamma\subset\partial\Omega$. Using geometric control, we could choose the form of $\frac{\partial v}{\partial t}(x,0)$ and design a probe function v satisfying (3.15) together with

$$v(x,t) = 0 \qquad \text{on } \partial\Omega\setminus\bar\Gamma,$$

so that we had

$$\int_0^T \int_\Gamma \frac{\partial p}{\partial \nu}(x,t)v(x,t)d\sigma(x)dt = \int_\Omega p_0(x)\frac{\partial v}{\partial t}(x,0)dx. \qquad (3.19)$$

Varying our choice of $\frac{\partial v}{\partial t}(x,0)$, we could adapt classical imaging algorithms (MUSIC, back-propagation, Kirchhoff migration, arrival-time) to the case of limited view data.

Now simply consider the 2-parameter family of probe functions:

$$v_{\tau,\theta}^{(2)}(x,t) = 1 - H\left(\frac{x\cdot\theta}{c_0}+t-\tau\right), \qquad (3.20)$$

where H is the Heaviside function. The probe function $v_{\tau,\theta}^{(2)}(x,t)$ is an incoming plane wavefront. Its equivalent, still denoted by $v_{\tau,\theta}^{(2)}$, in the limited-view setting satisfies the initial conditions

$$v_{\tau,\theta}^{(2)}(x,0) = 0 \quad \text{and} \quad \frac{\partial v_{\tau,\theta}^{(2)}}{\partial t}(x,0) = \delta\left(\frac{x\cdot\theta}{c_0}-\tau\right), \qquad (3.21)$$

together with the boundary condition $v_{\tau,\theta}^{(2)} = 0$ on $\partial\Omega \setminus \Gamma \times (0, T)$.

Note that if $T \geq \frac{\text{diam}(\Omega)}{c_0}$ in the full-view setting, our test functions $v_{\tau,\theta}^{(1)}$, $v_{\tau,\theta}^{(2)}$ and $w_{\tau,y}$ vanish at $t = T$. In the limited-view case, under the geometric controllability conditions [5] on Γ and T, existence of the test function v is guaranteed.

In both the full- and the limited-view cases, we get

$$\int_0^T \int_{\partial\Omega \text{ or } \Gamma} \frac{\partial p}{\partial \nu}(x, t) v_{\tau,\theta}^{(2)}(x, t) d\sigma(x) dt = \mathrm{R}[p_0](\theta, \tau), \qquad (3.22)$$

where $\mathrm{R}[f]$ is the (line) Radon transform of f. Applying a classical filtered back-projection algorithm to the data (3.22), one can reconstruct $p_0(x)$.

To illustrate the need of this approach, we present in Fig. 3.9 the reconstruction results from data with homogeneous Dirchlet boundary conditions. We compare the reconstruction using the inverse spherical Radon transform with the duality approach presented above. It appears that not taking boundary conditions into account leads to important errors in the reconstruction.

We then tested this approach on the Shepp-Logan phantom, using the family of probe functions $v_{\tau,\theta}^{(2)}$. Reconstructions are given in Fig. 3.10. We notice numerical noise due to the use of discontinuous (Heaviside) test functions against discrete measurements.

The numerical tests were conducted using Matlab. Three different forward solvers have been used for the wave equation:

- A FDTD solver, with Newmark scheme for time differentiation.
- A space-Fourier solver, with Crank-Nicholson finite difference scheme in time.
- a space-(P1) FEM-time finite difference solver.

Measurements were supposed to be obtained on equi-distributed captors on a circle or a square. The use of integral transforms (line or spherical Radon

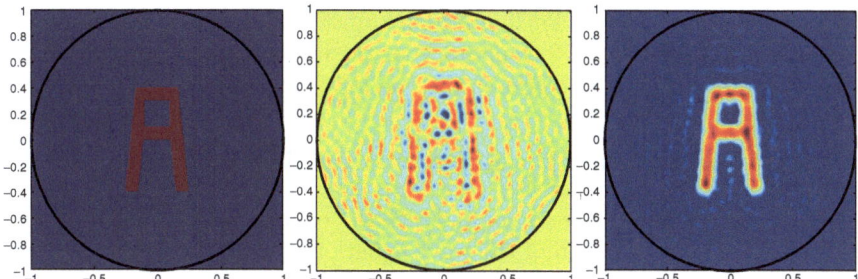

Fig. 3.9. Reconstruction in the case of homogeneous Dirichlet boundary conditions. *Left*: initial condition p_0; *Center*: reconstruction using spherical Radon transform; *Right*: reconstruction using probe functions algorithm

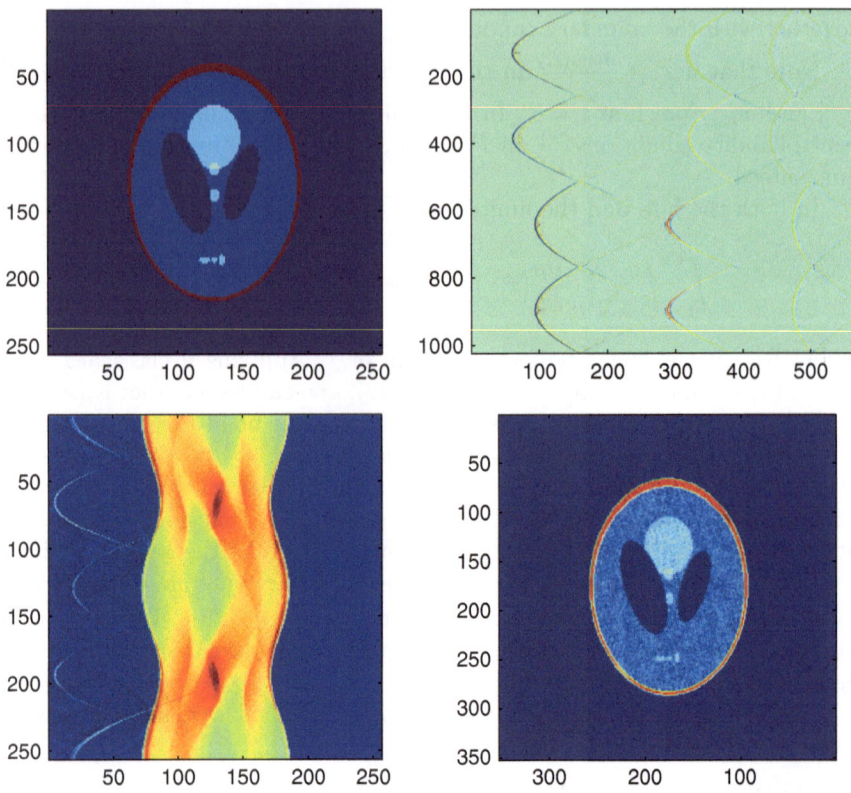

Fig. 3.10. Numerical inversion in the case of homogeneous Dirichlet boundary conditions. Here, $N = 256$, $N_R = 200$ and $N_\theta = 200$. *Top left*: p_0; *Top right*: $p(y,t)$ with $(y,t) \in \partial\Omega \times (0,3)$; *Bottom left*: $\mathcal{R}[p_0]$; *Bottom right*: reconstruction using probe functions algorithm

transform) avoids inverse crime since such transforms are computed on a different class of parameters (center and radius for spherical Radon transforms, direction and shift for line Radon transform). Indeed, their numerical inversions (achieved using formula (3.3) or the iradon function of Matlab) are not computed on the same grid as the one for the forward solvers.

3.3.3 Compensation of the Effect of Acoustic Attenuation

Our aim in this section is to compensate the effect of acoustic attenuation. Let $p_a(x,t)$ be the solution of the wave equation in a dissipative medium:

$$\frac{1}{c_0^2}\frac{\partial^2 p_a}{\partial t^2}(x,t) - \Delta p_a(x,t) - L(t) * p_a(x,t) = \frac{1}{c_0^2}\delta'_{t=0}p_0(x) \quad \text{in } \Omega \times \mathbb{R}, \quad (3.23)$$

with the Dirichlet (resp. the Neumann) imposed boundary conditions:

$$p_a(x,t) = 0 \quad \left(\text{resp. } \frac{\partial p_a}{\partial \nu}(x,t) = 0\right) \qquad \text{on } \partial\Omega \times \mathbb{R}, \tag{3.24}$$

where L is defined by (3.6).

We want to recover $p_0(x)$ from boundary measurements of $\dfrac{\partial p_a}{\partial \nu}(x,t)$ (resp. $p_a(x,t)$). Again, we assume that a is small.

Taking the Fourier transform of (3.23) yields

$$\begin{cases} (\Delta + K^2(\omega))\hat{p}_a(x,\omega) = \dfrac{i\omega}{\sqrt{2\pi}c_0^2}p_0(x) & \text{in } \Omega, \\[2ex] \hat{p}_a(x,\omega) = 0 \quad \left(\text{resp. } \dfrac{\partial \hat{p}_a}{\partial \nu}(x,\omega) = 0\right) & \text{on } \partial\Omega, \end{cases} \tag{3.25}$$

where $\hat{p}_a$ denotes the Fourier transform of p_a.

3.3.4 Case of a Spherical Wave as a Probe Function

By multiplying (3.25) by the Fourier transform, $\hat{w}_{0,y}(x,\omega)$, of $w_{\tau=0,y}$ given by (3.18), we arrive at, for any τ,

$$\frac{i}{\sqrt{2\pi}}\int_\Omega p_0(x)\left(\int_{\mathbb{R}} \omega e^{i\omega\tau}\hat{w}_{0,y}(x,K(\omega))\,d\omega\right)dx$$
$$= \int_{\mathbb{R}} e^{i\omega\tau}\int_{\partial\Omega} \frac{\partial \hat{p}_a}{\partial \nu}(x,\omega)\hat{w}_{0,y}(x,K(\omega))\,d\omega, \tag{3.26}$$

for the Dirichlet problem and

$$\frac{i}{\sqrt{2\pi}}\int_\Omega p_0(x)\left(\int_{\mathbb{R}} \omega e^{i\omega\tau}\hat{w}_{0,y}(x,K(\omega))\,d\omega\right)dx$$
$$= -\int_{\mathbb{R}} e^{i\omega\tau}\int_{\partial\Omega} \hat{p}_a(x,\omega)\frac{\partial \hat{w}_{0,y}}{\partial \nu}(x,K(\omega))\,d\omega, \tag{3.27}$$

for the Neumann problem.

Next we compute $\int_{\mathbb{R}} \omega e^{i\omega\tau}\hat{w}_{0,y}(x,K(\omega))\,d\omega$ for the thermo-viscous model. Recall that in this case,

$$K(\omega) \approx \frac{\omega}{c_0} + \frac{ia\omega^2}{2}.$$

We have

$$\int_{\mathbb{R}} \omega e^{i\omega\tau}\hat{w}_{0,y}(x,K(\omega))\,d\omega \approx \frac{1}{4\pi|x-y|}\int_{\mathbb{R}} \omega e^{i\omega(\tau-\frac{|x-y|}{c_0})}e^{-a\omega^2\frac{|x-y|}{c_0}}\,d\omega, \tag{3.28}$$

and again, the stationary phase theorem can then be applied to approximate the inversion procedure for $p_0(x)$.

Note that if we use the Fourier transform $\hat{v}$ of (3.17) or (3.20) as a test function then we have to truncate the integral in (3.26) since $\hat{v}(x,K(\omega))$ is exponentially growing in some regions of Ω.

3.3.5 Case of a Plane Wave as a Probe Function

Let us first introduce the function $\tilde{K}(\omega)$ defined by $\tilde{K}(\omega) = \sqrt{K(\omega)^2}$ and consider a solution of the Helmholtz equation

$$\left(\Delta + \tilde{K}^2(\omega) \right) \hat{v}_a(x, \omega) = 0$$

of the form

$$\hat{v}_a(x, \omega) = e^{-i\omega(x \cdot \theta - c_0 \tau)} g(\omega), \tag{3.29}$$

where $g(\omega)$ decays sufficiently fast.

Multiplying (3.25) by $\hat{v}_a(x, \omega)$, we obtain

$$\frac{i}{\sqrt{2\pi}} \int_\Omega p_0(x) \left(\int_{\mathbb{R}} \omega \overline{\hat{v}_a(x, \omega)} d\omega \right) dx = \int_{\mathbb{R}} \int_{\partial \Omega} \frac{\partial \hat{p}_a}{\partial \nu}(x, \omega) \overline{\hat{v}_a(x, \omega)} d\sigma(x) d\omega. \tag{3.30}$$

Since $\tilde{K}(\omega) \simeq \frac{\omega}{c_0} - \frac{ia\omega^2}{2}$, then by taking in formula (3.29)

$$g(\omega) = e^{-\frac{1}{2}\omega^2 a c_0 T} \quad \text{and} \quad g(\omega) = \frac{1}{i\omega} e^{-\frac{1}{2}\omega^2 a c_0 T},$$

we can use the plane waves $\hat{v}_a^{(1)}$ and $\hat{v}_a^{(2)}$ given by

$$\hat{v}_a^{(1)}(x, \omega) = e^{-i\omega(x \cdot \theta - c_0 \tau)} e^{-\frac{1}{2}\omega^2 a c_0 (T + \frac{x \cdot \theta}{c_0} - \tau)},$$

and

$$\hat{v}_a^{(2)}(x, \omega) = \frac{1}{i\omega} e^{-i\omega(x \cdot \theta - c_0 \tau)} e^{-\frac{1}{2}\omega^2 a c_0 (T + \frac{x \cdot \theta}{c_0} - \tau)},$$

as approximate probe functions.

Take T sufficiently large such that $\left(T + \frac{x \cdot \theta}{c_0} - \tau \right)$ stays positive for all $x \in \Omega$. Thus,

$$v_a^{(1)}(x, t) \simeq \frac{1}{\sqrt{ac_0 \left(T + \frac{x \cdot \theta}{c_0} - \tau \right)}} e^{-\dfrac{(x \cdot \theta - c_0 \tau + t)^2}{2ac_0 \left(T + \frac{x \cdot \theta}{c_0} - \tau \right)}},$$

and

$$v_a^{(2)}(x, t) \simeq \text{erf}\left(\frac{x \cdot \theta - c_0 \tau + t}{\sqrt{ac_0 \left(T + \frac{x \cdot \theta}{c_0} - \tau \right)}} \right).$$

Now using $v_a^{(2)}$ in formula (3.30) leads to the convolution of the Radon transform of p_0 with a quasi-Gaussian kernel. Indeed, the left hand-side of (3.30) satisfies

$$\frac{i}{\sqrt{2\pi}} \int_\Omega p_0(x) \left(\int_\mathbb{R} \omega \overline{\hat{v}_a^{(2)}(x, w)} dw \right) dx$$

$$\simeq \int_\Omega p_0(x) \frac{1}{\sqrt{ac_0 \left(T + \frac{x \cdot \theta}{c_0} - \tau\right)}} e^{-\frac{(x \cdot \theta - c_0\tau)^2}{2ac_0 \left(T + \frac{x \cdot \theta}{c_0} - \tau\right)}} dx$$

$$= \int_{s_{min}}^{s_{max}} R[p_0](\theta, s) \frac{1}{\sqrt{ac_0 \left(T + \frac{s}{c_0} - \tau\right)}} e^{-\frac{(s - c_0\tau)^2}{2ac_0 \left(T + \frac{s}{c_0} - \tau\right)}} ds,$$

and the right hand-side is explicitly estimated by

$$\int_\mathbb{R} \int_{\partial\Omega} \frac{\partial \hat{p}_a}{\partial \nu}(x, \omega) \overline{\hat{v}_a^{(2)}(x, \omega)} d\sigma(x) d\omega$$

$$\simeq \int_0^T \int_{\partial\Omega} \frac{\partial p_a}{\partial \nu}(x, t) \mathrm{erf}\left(\frac{x \cdot \theta - c_0\tau + t}{\sqrt{ac_0 \left(T + \frac{x \cdot \theta}{c_0} - \tau\right)}}\right) d\sigma(x) dt.$$

As previously, we can compensate the effect of attenuation using the stationary phase theorem for the operator $\tilde{\mathcal{L}}$,

$$\tilde{\mathcal{L}}[\phi](\tau) = \int_{s_{min}}^{s_{max}} \phi(s) \frac{1}{\sqrt{ac_0 \left(T + \frac{s}{c_0} - \tau\right)}} e^{-\frac{(s - c_0\tau)^2}{2ac_0 \left(T + \frac{s}{c_0} - \tau\right)}} ds,$$

which reads

$$\tilde{\mathcal{L}}[\phi](\tau) \simeq \phi(c_0\tau) + \frac{ac_0 T}{2} \left(\phi''(c_0\tau) + \frac{2\phi'(c_0\tau)}{c_0 T}\right). \tag{3.31}$$

See Appendix 3. More generally,

$$\tilde{\mathcal{L}}[\phi](\tau) = \sum_{i=0}^k \frac{(c_0 a)^i}{2^i i!} D_i[\phi] + o(a^k), \tag{3.32}$$

where the differential operators D_i satisfy

$$D_i[\phi] = ((T + \frac{s}{c_0} - \tau)^i [\phi](s))_{|s=c_0\tau}^{(2i)}.$$

Define $\tilde{\mathcal{L}}_k^{-1}$ as in (3.11). Using (3.32), we reconstructed the line Radon transform of p_0 correcting the effect of attenuation. We then applied a standard filtered back-projection algorithm to inverse the Radon transform. Results are given in Fig. 3.11.

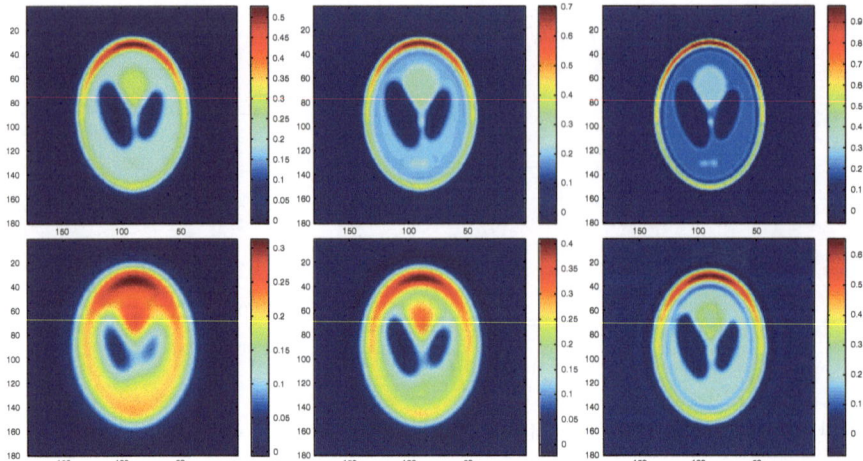

Fig. 3.11. Compensation of acoustic attenuation with formula (3.31) in the case of homogeneous Dirichlet boundary conditions. Here, $N = 256$, $N_R = 200$ and $N_\theta = 200$. *First line*: $a = 0.0005$; *Second line*: $a = 0.0025$. *Left*: $\tilde{\mathcal{L}}_k^{-1}$ with $k = 0$; *Center*: $\tilde{\mathcal{L}}_k^{-1}$ with $k = 1$; *Right*: $\tilde{\mathcal{L}}_k^{-1}$ with $k = 8$

3.4 Conclusion

In this chapter we have provided new approaches to correct the effect of imposed boundary conditions as well as the effect of acoustic attenuation.

It would be very interesting to analytically investigate their robustness with respect to measurement noise and medium noise. In this connection, we refer to [4] for a coherent interferometric (CINT) strategy for photo-acoustic imaging in the presence of microscopic random fluctuations of the speed of sound.

Another important problem is to *a priori* estimate the attenuation coefficient a and the frequency power ζ.

Finally, it is worth emphasizing that it is the absorption coefficient, not the absorbed energy, that is a fundamental physiological parameter. The absorbed energy density is in fact the product of the optical absorption coefficient and the light fluence which depends on the distribution of scattering and absorption within the domain, as well as the light sources. In [3], methods for reconstructing the normalized optical absorption coefficient of small absorbers from the absorbed density are proposed. Multi-wavelength acoustic measurements are combined with diffusing light measurements to separate the product of absorption coefficient and optical fluence. In the case of extended absorbers, multi-wavelength photo-acoustic imaging is also expected to lead to a satisfactory solution [11]. See also Chap. 5, where an efficient algorithm for large-scale three-dimensional quantitative photo-acoustic reconstructions is proposed.

Appendix 1: Stationary Phase Theorem

Theorem 3.1 (Stationary Phase [18]) *Let $K \subset [0, \infty)$ be a compact set, X an open neighborhood of K and k a positive integer. If $\psi \in C_0^{2k}(K)$, $f \in C^{3k+1}(X)$ and $Im(f) \geq 0$ in X, $Im(f(t_0)) = 0$, $f'(t_0) = 0$, $f''(t_0) \neq 0$, $f' \neq 0$ in $K \setminus \{t_0\}$ then for $\epsilon > 0$*

$$\left| \int_K \psi(t) e^{if(t)/\epsilon} dt - e^{if(t_0)/\epsilon} \left(\epsilon^{-1} f''(t_0)/2\pi i \right)^{-1/2} \sum_{j<k} \epsilon^j L_j[\psi] \right|$$

$$\leq C\epsilon^k \sum_{\alpha \leq 2k} \sup |\psi^{(\alpha)}(x)|.$$

Here C is bounded when f stays in a bounded set in $C^{3k+1}(X)$ and $|t - t_0|/|f'(t)|$ has a uniform bound. With,

$$g_{t_0}(t) = f(t) - f(t_0) - \frac{1}{2} f''(t_0)(t - t_0)^2,$$

which vanishes up to third order at t_0, and

$$L_j[\psi] = \sum_{\nu - \mu = j} \sum_{2\nu \geq 3\mu} i^{-j} \frac{2^{-\nu}}{\nu!\mu!} (-1)^\nu f''(t_0)^{-\nu} (g_{t_0}^\mu \psi)^{(2\nu)}(t_0).$$

We will use this theorem with $k = 2$. Note that L_1 can be expressed as the sum $L_1[\psi] = L_1^{(1)}[\psi] + L_1^{(2)}[\psi] + L_1^{(3)}[\psi]$, where $L_1^{(j)}$ is respectively associated to the couple $(\nu_j, \mu_j) = (1, 0), (2, 1), (3, 2)$ and is identified as

$$\begin{cases} L_1^{(1)}[\psi] &= -\frac{1}{2i} f''(t_0)^{-1} \psi^{(2)}(t_0), \\ L_1^{(2)}[\psi] &= \frac{1}{2^2 2! i} f''(t_0)^{-2} (g_{t_0}\psi)^{(4)}(t_0) \\ &= \frac{1}{8i} f''(t_0)^{-2} \left(g_{t_0}^{(4)}(t_0)\psi(t_0) + 4g_{t_0}^{(3)}(t_0)\psi'(t_0) \right), \\ \nu L_1^{(3)}[\psi] &= \frac{-1}{2^3 2! 3! i} f''(t_0)^{-3} (g_{t_0}^2 \psi)^{(6)}(t_0) = \frac{-1}{2^3 2! 3! i} f''(t_0)^{-3} (g_{t_0}^2)^{(6)}(t_0)\psi(t_0). \end{cases}$$

Appendix 2: Proof of Approximation (3.10)

Let us now apply the stationary phase theorem to the operator $\tilde{\mathcal{L}}$

$$\tilde{\mathcal{L}}[\phi] = \frac{1}{\sqrt{2\pi}} \int_0^{+\infty} \phi(t) \frac{1}{\sqrt{c_0 a t}} e^{-\frac{1}{2} \frac{(s-t)^2}{c_0 a t}} dt.$$

Note that the integral

$$J(s) = \int_0^\infty \psi(t) e^{if(t)/\epsilon} dt,$$

with $\psi(t) = \frac{\phi(t)}{\sqrt{t}}$, $\epsilon = c_0 a$, $f(t) = i\frac{(t-s)^2}{2t}$, satisfies $J(s) = \sqrt{c_0 a 2\pi}\tilde{\mathcal{L}}[\phi]$. The phase f vanishes at $t = s$ and satisfies

$$f'(t) = i\frac{1}{2}\left(1 - \frac{s^2}{t^2}\right), \quad f''(t) = i\frac{s^2}{t^3}, \quad f''(s) = i\frac{1}{s}.$$

The function $g_s(t)$ is given by

$$g_s(t) = i\frac{1}{2}\frac{(t-s)^2}{t} - i\frac{1}{2}\frac{(t-s)^2}{s} = i\frac{1}{2}\frac{(s-t)^3}{ts}.$$

We can deduce that

$$\begin{cases} (g_s\psi)^{(4)}(s) = \left(g_{x_0}^{(4)}(s)\psi(s) + 4g_{x_0}^{(3)}(s)\psi'(s)\right) = i\frac{1}{2}\left(\frac{24}{s^3}\psi(s) - \frac{24}{s^2}\psi'(s)\right), \\ (g_s^2\psi)^{(6)}(s) = (g_{x_0}^2)^{(6)}(s)\psi(s) = -\frac{1}{4}\frac{6!}{s^4}\psi(s), \end{cases}$$

and then, with the same notation as in Theorem 3.1,

$$\begin{cases} L_1^{(1)}[\psi] = -\frac{1}{i}\left(\frac{1}{2}(f''(s))^{-1}\psi''(s)\right) = \frac{1}{2}s\left(\frac{\phi}{\sqrt{s}}\right)'' = \frac{1}{2}\left(\sqrt{s}\phi''(s) - \frac{\phi'(s)}{\sqrt{s}} + \frac{3}{4}\frac{\phi}{s^{3/2}}\right), \\ L_1^{(2)}[\psi] = \frac{1}{8i}f''(s)^{-2}\left(g_s^{(4)}(s)\psi(s) + 4g_s^{(3)}(s)\psi'(s)\right) = \frac{1}{2}\left(3\left(\frac{\phi(s)}{\sqrt{s}}\right)' - 3\frac{\phi(s)}{s^{3/2}}\right) \\ \qquad = \frac{1}{2}\left(3\frac{\phi'(s)}{\sqrt{s}} - \frac{9}{2}\frac{\phi(s)}{s^{3/2}}\right), \\ L_1^{(3)}[\psi] = \frac{-1}{2^3 2! 3! i}f''(s)^{-3}(g_s^2)^{(6)}(s)\psi(s) = \frac{1}{2}\left(\frac{15}{4}\frac{\phi(s)}{s^{3/2}}\right). \end{cases}$$

The operator L_1 is given by

$$\begin{aligned} L_1[\psi] &= L_1^{(1)}[\psi] + L_1^{(2)}[\psi] + L_1^{(3)}[\psi] \\ &= \frac{1}{2}\left(\sqrt{s}\phi''(s) + (3-1)\frac{\phi'(s)}{\sqrt{s}} + \left(\frac{3}{4} - \frac{9}{2} + \frac{15}{4}\right)\frac{\phi(s)}{s^{3/2}}\right) \\ &= \frac{1}{2\sqrt{s}}(s\phi(s))'', \end{aligned}$$

and so,

$$\left| J(s) - \sqrt{2\pi a c_0 s}\left(\frac{\phi(s)}{\sqrt{s}} + a\frac{1}{2\sqrt{s}}(s\phi(s))''\right)\right| \leq Ca^2\sum_{\alpha\leq 4}\sup|\phi^{(\alpha)}(x)|.$$

Finally, we arrive at

$$\left|\frac{1}{\sqrt{2\pi}}\int_0^\infty \phi(t)\frac{1}{\sqrt{ac_0 t}}e^{-\frac{(t-s)^2}{2ac_0 t}}dt - \left(\phi(s) + \frac{a}{2}(s\phi(s))''\right)\right|$$
$$\leq Ca^{3/2}\sum_{\alpha\leq 4}\sup|\phi^{(\alpha)}(t)|.$$

Acknowledgments

This work was supported by the ERC Advanced Grant Project MULTIMOD–267184.

Appendix 3: Proof of Approximation (3.31)

Let us now apply the stationary phase theorem to the operator $\mathcal{L}$ defined by

$$\tilde{\mathcal{L}}[\phi](\tau) = \frac{1}{\sqrt{2\pi}} \int_{s_{\min}}^{s_{\max}} \left[\phi(s) \left(a \left(c_0 T + s - c_0 \tau \right) \right)^{-\frac{1}{2}} e^{-\frac{(s - c_0\tau)^2}{2a\left(c_0 T + s - c_0\tau\right)}} \right] ds$$

$$= \frac{1}{\sqrt{2\pi}} \int_{s_{\min}-c_0\tau}^{s_{\max}-c_0\tau} \left[\phi(t + c_0\tau) \left(a\left(\tilde{T}+t\right) \right)^{-\frac{1}{2}} e^{-\frac{t^2}{2a\left(\tilde{T}+t\right)}} \right] dt,$$

where $\tilde{T} = c_0 T$. Note that the integral

$$J(\tau) = \int_{s_{\min}-c_0\tau}^{s_{\max}-c_0\tau} \psi(t) e^{if(t)/\epsilon} dt,$$

with $\psi(t) = \frac{\phi(t+c_0\tau)}{\sqrt{\tilde{T}+t}}$, $\epsilon = a$, $f(s) = i\frac{t^2}{2(\tilde{T}+t)}$, satisfies $J(\tau) = \sqrt{a2\pi}\tilde{\mathcal{L}}[\phi]$.
The phase f vanishes at $t = 0$ and satisfies

$$f'(t) = i\frac{1}{2}\frac{t(t + 2\tilde{T})}{(t + \tilde{T})^2}, \quad f''(t) = i\frac{\tilde{T}^2}{(t + \tilde{T})^3}, \quad f''(0) = i\frac{1}{\tilde{T}}.$$

The function $g_0(t)$ is identified as

$$g_0(t) = -i\frac{1}{2}\frac{t^3}{\tilde{T}(\tilde{T} + t)}.$$

We have

$$\begin{cases} (g_0\psi)^{(4)}(0) &= \left(g_0^{(4)}(0)\psi(0) + 4g_0^{(3)}(0)\psi'(0) \right) = i\frac{1}{2}\left(\frac{24}{\tilde{T}^3}\psi(0) - \frac{24}{\tilde{T}^2}\psi'(0) \right), \\ (g_0^2\psi)^{(6)}(0) &= (g_0^2)^{(6)}(0)\psi(0) = -\frac{1}{4}\frac{6!}{\tilde{T}^4}\psi(0), \end{cases}$$

and

$$\psi(0) = \frac{\phi(c_0\tau)}{\tilde{T}^{1/2}}, \quad \psi'(0) = \frac{\phi'(c_0\tau)}{\tilde{T}^{1/2}} - \frac{1}{2}\frac{\phi(c_0\tau)}{\tilde{T}^{3/2}},$$

$$\psi''(0) = \frac{\phi''(c_0\tau)}{\tilde{T}^{1/2}} - \frac{\phi'(c_0\tau)}{\tilde{T}^{3/2}} + \frac{3}{4}\frac{\phi(c_0\tau)}{\tilde{T}^{5/2}}.$$

Therefore, again with the same notation as in Theorem 3.1,

$$
\begin{cases}
L_1^{(1)}[\psi] = -\dfrac{1}{i}\left(\dfrac{1}{2}(f''(0))^{-1}\psi''(0)\right) = \dfrac{1}{2}\left(\sqrt{\tilde{T}}\phi''(c_0\tau) - \dfrac{\phi'(c_0\tau)}{\tilde{T}^{1/2}} + \dfrac{3}{4}\dfrac{\phi(c_0\tau)}{\tilde{T}^{3/2}}\right), \\[3mm]
L_1^{(2)}[\psi] = \dfrac{1}{8i}f''(0)^{-2}\left(g_0^{(4)}(0)\psi(0) + 4g_0^{(3)}(0)\psi'(0)\right) = \dfrac{1}{2}\left(3\psi'(0) - 3\dfrac{\psi(0)}{\tilde{T}}\right) \\[3mm]
\qquad = \dfrac{1}{2}\left(3\dfrac{\phi'(c_0\tau)}{\tilde{T}^{1/2}} - \dfrac{9}{2}\dfrac{\phi(c_0\tau)}{\tilde{T}^{3/2}}\right), \\[3mm]
L_1^{(3)}[\psi] = -\dfrac{1}{2^3 2! 3! i}f''(0)^{-3}(g_0^2)^{(6)}(0)\psi(0) = \dfrac{1}{2}\left(\dfrac{15}{4}\dfrac{\phi(c_0\tau)}{\tilde{T}^{3/2}}\right),
\end{cases}
$$

and L_1 is given by

$$
L_1[\psi] = L_1^{(1)}[\psi] + L_1^{(2)}[\psi] + L_1^{(3)}[\psi] = \dfrac{1}{2\sqrt{\tilde{T}}}\left(\tilde{T}\phi''(c_0\tau) + 2\phi'(c_0\tau)\right)
$$

$$
= \dfrac{1}{2\sqrt{\tilde{T}}}\left((s - c_0\tau + \tilde{T})\phi(s)\right)''_{\lfloor s=c_0\tau},
$$

which yields

$$
\left| J(\tau) - \sqrt{2\pi a}\left(\phi(c_0\tau) + a/2\left((s - c_0\tau + c_0T)\phi(s)\right)''_{s=c_0\tau}\right)\right|
$$
$$
\leq Ca^2 \sum_{\alpha \leq 4} \sup |\phi^{(\alpha)}(x)|.
$$

Hence,

$$
\left|\tilde{\mathcal{L}}[\phi] - \left(\phi(c_0\tau) + \dfrac{ac_0T}{2}\left(\phi''(c_0\tau) + \dfrac{2\phi'(c_0\tau)}{c_0T}\right)\right)\right| \leq Ca^{3/2} \sum_{\alpha \leq 4} \sup |\phi^{(\alpha)}(t)|.
$$

References

1. H. Ammari, M. Asch, L. Guadarrama Bustos, V. Jugnon, H. Kang, Transient wave imaging with limited-view data, SIAM J. Imag. Sci, to appear.
2. H. Ammari, E. Bossy, V. Jugnon, H. Kang, Mathematical modelling in photo-acoustic imaging of small absorbers, SIAM Rev. **52**, 677–695 (2010)
3. H. Ammari, E. Bossy, V. Jugnon, H. Kang, Quantitative photo-acoustic imaging of small absorbers, SIAM J. Appl. Math. **71**, 676–693 (2011)
4. H. Ammari, E. Bretin, J. Garnier, V. Jugnon, Coherent interferometric algorithms for photoacoustic imaging, SIAM J. Numer. Anal., to appear.
5. C. Bardos, G. Lebeau, J. Rauch, Sharp sufficient conditions for the observation, control, and stabilization of waves from the boundary, SIAM J. Contr. Optim. **30**, 1024–1065 (1992)
6. A. Beck, M. Teboulle, A fast iterative shrinkage-thresholding algorithm for linear inverse problems, SIAM J. Imaging Sci. **2**, 183–202 (2009)

7. P. Burgholzer, H. Grün, M. Haltmeier, R. Nuster, G. Paltauf, Compensation of acoustic attenuation for high resolution photoacoustic imaging with line detectors, Proc. of SPIE, **6437**, 643724 (2007)

8. A. Chambolle, An algorithm for total variation minimization and applications, J. Math. Imag. Vis. **20**, 89–97 (2004)

9. W. Chen, S. Holm, Fractional Laplacian time-space models for linear and nonlinear lossy media exhibiting arbitrary frequency power-law dependency, J. Acoust. Soc. Amer. **115**, 1424–1430 (2004)

10. P. Combettes, V. Wajs, Signal recovery by proximal forward-backward splitting, Mult. Model. Simul. **4**, 1168–1200 (2005)

11. B.T. Cox, J.G. Laufer, P.C. Beard, The challenges for quantitative photoacoustic imaging, Proc. of SPIE **7177**, 717713 (2009)

12. I. Daubechies, M. Defrise, C. De Mol, An iterative thresholding algorithm for linear inverse problems with a sparsity constraint, Comm. Pure Appl. Math. **57**, 1413–1457 (2004)

13. D. Finch, M. Haltmeier, Rakesh, Inversion of spherical means and the wave equation in even dimensions, SIAM J. Appl. Math. **68**, 392–412 (2007)

14. D. Finch, S. Patch, Rakesh, Determining a function from its mean-values over a family of spheres, SIAM J. Math. Anal. **35**, 1213–1240 (2004)

15. M. Haltmeier, R. Kowar, A. Leitao, O. Scherzer, Kaczmarz methods for regularizing nonlinear ill-posed equations II: Applications, Inverse Probl. Imag. **1**, 507–523 (2007)

16. M. Haltmeier, T. Schuster, O. Scherzer, Filtered backprojection for thermoacoustic computed tomography in spherical geometry, Math. Meth. Appl. Sci. **28**, 1919-1937 (2005)

17. M. Haltmeier, O. Scherzer, P. Burgholzer, R. Nuster, G. Paltauf, Thermoacoustic tomography and the circular Radon transform: exact inversion formula, Math. Model. Meth. Appl. Sci. **17**(4), 635-655 (2007)

18. L. Hörmander, The Analysis of Linear Partial Differential Operators. I. Distribution Theory and Fourier Analysis, Classics in Mathematics, Springer, Berlin (2003)

19. R. Kowar, Integral equation models for thermoacoustic imaging of dissipative tissue, Inverse Probl. **26**, 095005 (18pp) (2010)

20. R. Kowar, O. Scherzer, X. Bonnefond, Causality analysis of frequency dependent wave attenuation, Math. Meth. Appl. Sci. **34**, 108–124 (2011)

21. P. Kuchment, L. Kunyansky, in *Mathematics of Photoacoustics and Thermoacoustic Tomography*, ed. by O. Scherzer. Handbook of Mathematical Methods in Imaging (Springer, New York, 2011)

22. P. Kunchment, L. Kunyansky, Mathematics of thermoacoustic tomography, Europ. J. Appl. Math. **19**, 191–224 (2008)

23. L. Kunyansky, Explicit inversion formulas for the spherical mean Radon transform, Inverse Probl. **23**, 373–383 (2007)

24. P.J. La Rivière, J. Zhang, M.A. Anastasio, Image reconstruction in optoacoustic tomography for dispersive acoustic media, Opt. Lett. **31**, 781–783 (2006)

25. K. Maslov, H.F. Zhang, L.V. Wang, Effects of wavelength-dependent fluence attenuation on the noninvasive photoacoustic imaging of hemoglobin oxygen saturation in subcutaneous vasculature in vivo, Inverse Probl. **23**, S113–S122 (2007)

26. L.V. Nguyen, A family of inversion formulas in thermoacoustic tomography, Inverse Probl. Imag. **3**, 649–675 (2009)

27. K. Patch, M. Haltmeier, Thermoacoustic tomography – ultrasound attenuation artifacts, IEEE Nucl. Sci. Symp. Conf. **4**, 2604–2606 (2006)
28. N.V. Sushilov, R.S.C. Cobbold, Frequency-domain wave equation and its time-domain solutions in attenuating media, J. Acoust. Soc. Am. **115**, 1431–1436 (2004)
29. T.L. Szabo. Causal theories and data for acoustic attenuation obeying a frequency power law. J. Acoust. Soc. Am. **97**, 14–24 (1995)
30. B.E. Treeby, B.T. Cox, Fast, tissue-realistic models of photoacoustic wave propagation for homogeneous attenuating media, Proc. of SPIE **7177**, 717716 (2009)
31. L.V. Wang, X. Yang, Boundary conditions in photoacoustic tomography and image reconstruction, J. Biomed. Optic. **12**, 014027 (2007)
32. M. Xu, L.V. Wang, Photoacoustic imaging in biomedicine, Rev. Scient. Instrum. **77**, 041101 (2006)
33. Y. Xu, L.V. Wang, Reconstructions in limited-view thermoacoustic tomography, Med. Phys. **31**, 724–733 (2004)

4

Attenuation Models in Photoacoustics

Richard Kowar[1] and Otmar Scherzer[*,2]

[1]Institute of Mathematics, University of Innsbruck, Technikerstr. 21a,
6020 Innsbruck, Austria `richard.kowar@uibk.ac.at`
[2]Computational Science Center, University of Vienna, Nordbergstr. 15,
1090 Vienna, Austria `otmar.scherzer@univie.ac.at`

Summary. The aim of this chapter is to review attenuation models in photo-acoustic imaging and discuss their causality properties. We also derive integro-differential equations which the attenuated waves are satisfying and highlight the ill–conditionness of the inverse problem for calculating the unattenuated wave from the attenuated one, which has been discussed in Chap. 3.

4.1 Introduction

Photoacoustic Imaging is one of the recent hybrid imaging techniques, which attempts to visualize the distribution of the *electromagnetic absorption coefficient* inside a biological object. In photoacoustic experiments, the medium is exposed to a short pulse of a relatively low frequency electromagnetic (EM) wave. The exposed medium absorbs a fraction of the EM energy, heats up, and reacts with thermoelastic expansion. This induces acoustic waves, which can be recorded outside the object and used to determine the electromagnetic absorption coefficient. The combination of EM and ultrasound waves (which explains the usage of the term *hybrid*) allows one to combine high contrast in the EM absorption coefficient with high resolution of ultrasound. The method has demonstrated great potential for biomedical applications, including functional brain imaging of animals [47], soft-tissue characterization, and early stage cancer diagnostics [23], as well as imaging of vasculature [55]. For a general survey on biomedical applications see [53]. In comparison with the X-Ray CT, photoacoustics is non-ionizing. Its further advantage is that soft biological tissues display high contrasts in their ability to absorb frequency electromagnetic waves. For instance, for radiation in the near infrared domain, as produced by a Nd:YAG laser, the absorption coefficient in human soft tissues varies in the range of 0.1–$0.5 \, \mathrm{cm}^{-1}$ [7]. The contrast is also known to be high between healthy and cancerous cells, which makes photoacoustics a promising early cancer detection technique. Another application arises in biology: Multispectral optoacoustic tomography technique is capable of high-resolution

H. Ammari (ed.), *Mathematical Modeling in Biomedical Imaging II*,
Lecture Notes in Mathematics 2035, DOI 10.1007/978-3-642-22990-9_4,
© Springer-Verlag Berlin Heidelberg 2012

visualization of fluorescent proteins deep within highly light-scattering living organisms [35]. In contrast, the current fluorescence microscopy techniques are limited to the depth of several hundred micrometers, due to intense light scattering.

Different terms are often used to indicate different excitation sources: *Optoacoustics* refers to illumination in the visible light spectrum, *Photoacoustics* is associated with excitations in the visible and infrared range, and *Thermoacoustics* corresponds to excitations in the microwave or radio-frequency range. In fact, the carrier frequency of the illuminating pulse is varying, which is usually not taken into account in mathematical modeling. Since the corresponding mathematical models are equivalent, in the mathematics literature, the terms opto-, photo-, and thermoacoustics are used interchangeably. In this article, we are addressing only the *photoacoustic tomographic technique* PAT (which is mathematically equivalent to the thermoacoustic tomography TAT).

Various kinds of photoacoustic imaging techniques have been implemented. See, for instance, [3, 11, 13–15, 24–27, 30, 31, 37, 38, 41, 46, 48–51]. One should distinguish between photoacoustic *microscopy* (PAM) and *tomography* (PAT). In microscopy, the object is scanned pixel by pixel (or voxel by voxel). The measured pressure data provides an image of the electromagnetic absorption coefficient [56]. Tomography, on the other hand, measures pressure waves with detectors surrounding completely or partially the object. Then the internal distribution of the absorption coefficients is reconstructed using mathematical inversion techniques (see the sections below).

The common underlying mathematical equation of PAT is the *wave equation* for the pressure

$$\frac{1}{c_0^2}\frac{\partial^2 p}{\partial t^2}(\mathbf{x},t) - \nabla^2 p(\mathbf{x},t) = \frac{dj}{dt}(t)\left(\frac{\mu_{\text{abs}}(\mathbf{x})\beta(\mathbf{x})J(\mathbf{x})}{c_p(\mathbf{x})}\right)\,,\quad \mathbf{x}\in\mathbb{R}^3,\,t>0\,. \quad (4.1)$$

Here c_p denotes the specific heat capacity, J is the spatial intensity distribution, μ_{abs} denotes the absorption coefficient, β denotes the thermal expansion coefficient and c_0 denotes the speed of sound, which is commonly assumed to be constant. The assumption that there is no acoustic pressure before the object is illuminated at time $t = 0$ is expressed by

$$p(\mathbf{x},t) = 0\,,\qquad \mathbf{x}\in\mathbb{R}^3, t<0\,. \quad (4.2)$$

In PAT, $j(t)$ approximates a pulse, and can be considered as a δ-impulse $\delta(t)$. Introducing the shorthand notations

$$\rho(\mathbf{x}) := \frac{\mu_{\text{abs}}(\mathbf{x})\beta(\mathbf{x})J(\mathbf{x})}{c_p(\mathbf{x})}\,, \quad (4.3)$$

one reduces (4.1) and (4.2) to

$$\frac{1}{c_0^2}\frac{\partial^2 p}{\partial t^2}(\mathbf{x},t) - \nabla^2 p(\mathbf{x},t) = 0\,,\quad \mathbf{x}\in\mathbb{R}^3, t>0\,, \quad (4.4)$$

with initial values

$$p(\mathbf{x}, 0) = \rho(\mathbf{x}) , \quad \frac{\partial p}{\partial t}(\mathbf{x}, 0) = 0 \quad \mathbf{x} \in \mathbb{R}^3 . \tag{4.5}$$

The quantity ρ in (4.1) and (4.3) is a combination of several physical parameters. All along this chapter ρ should not be confused with the source term

$$f(\mathbf{x}, t) = \frac{dj}{dt}(t)\rho(\mathbf{x}) , \quad \mathbf{x} \in \mathbb{R}^3, t > 0 . \tag{4.6}$$

In PAT, some data about the pressure $p(\mathbf{x}, t)$ are measured and the main task is to reconstruct the initial pressure ρ from these data. While the excitation principle is always as described above and thus (4.4) holds, the specific type of data measured depends on the type of transducers used, and thus influences the mathematical model.

Nowadays there is a trend to incorporate more and more modeling into photoacoustic. In particular, taking into account locally varying *wave speed* and *attenuation*. Even more there is a novel trend to *qualitative photoacoustics*, which is concerned with estimating physical parameters from the imaging parameter of standard photoacoustics. In this chapter we focus on attenuation correction, where we survey some recent progress. Inversion with varying wave speed has been considered for instance in [1, 18], and is not further discussed here.

The outline of this chapter is as follows: First, we review existing attenuation models and discuss their causality properties, which we believe to be essential for algorithms for inversion with attenuated data; see Chap. 3. Then, we survey causality properties of common attenuation models. We also derive integro-differential equations which the attenuated waves are satisfying. In addition we highlight the ill–conditionness of the inverse problem discussed in Chap. 3 for calculating the unattenuated wave from the attenuated one.

4.2 Attenuation

The difficult issue of effects of and corrections for the attenuation of acoustic waves in PAT has been studied [4, 22, 32, 36], although no complete conclusion on the feasibility of these models has been reached.

Mathematical models for describing attenuation are formulated in the frequency domain, taking into account that attenuation disperses high frequency components more rapidly over traveled distance. Let $\mathcal{G}(\mathbf{x}, t)$ denote the attenuated wave which originates from an impulse ($\delta_{\mathbf{x},t}$-distribution) at $\mathbf{x} = 0$ at time $t = 0$. In mathematical terms $\mathcal{G}$ is the Green-function of attenuated wave equation. Moreover, we denote by

$$\mathcal{G}_0(\mathbf{x}, t) = \frac{\delta\left(t - \frac{|\mathbf{x}|}{c_0}\right)}{4\pi |\mathbf{x}|} \tag{4.7}$$

the Green function of the unattenuated wave equation; that is, it is the solution of (4.4), (4.5) with constant sound speed $c(x) \equiv c_0$ and initial conditions [8,19]

$$\mathcal{G}_0(\mathbf{x}, 0) = 0 \quad \text{and} \quad \frac{\partial \mathcal{G}_0}{\partial t}(\mathbf{x}, 0) = \delta_{\mathbf{x}, t}.$$

Common mathematical formulations of *attenuation* assume that

$$\mathcal{F}\{\mathcal{G}\}(\mathbf{x}, \omega) = \exp\left(-\beta^*(|\mathbf{x}|, \omega)\right) \mathcal{F}\{\mathcal{G}_0\}(\mathbf{x}, \omega), \quad \mathbf{x} \in \mathbb{R}^3, \omega \in \mathbb{R}. \quad (4.8)$$

Here $\mathcal{F}\{\cdot\}$ denotes the Fourier transform with respect to time t (cf. Appendix 4.8). Applying the inverse Fourier transform $\mathcal{F}^{-1}\{\cdot\}$ to (4.8) gives

$$\mathcal{G}(\mathbf{x}, t) = K(\mathbf{x}, t) *_t \mathcal{G}_0(\mathbf{x}, t) \qquad (*_t \text{ time convolution}) \qquad (4.9)$$

where

$$K(\mathbf{x}, t) := \frac{1}{\sqrt{2\pi}} \mathcal{F}^{-1}\left\{\exp\left(-\beta^*(|\mathbf{x}|, \cdot)\right)\right\}(t). \qquad (4.10)$$

From (4.9) and (4.7) it follows that

$$\begin{aligned}
\mathcal{G}(\mathbf{x}, t) &= K(\mathbf{x}, t) *_t \mathcal{G}_0(\mathbf{x}, t) \\
&= \int_{\mathbb{R}} K(\mathbf{x}, t - \tau) \frac{\delta(\tau - \frac{|\mathbf{x}|}{c_0})}{4\pi |\mathbf{x}|} d\tau \\
&= \frac{K\left(\mathbf{x}, t - \frac{|\mathbf{x}|}{c_0}\right)}{4\pi |\mathbf{x}|}.
\end{aligned}$$

Consequently,

$$\mathcal{G}(\mathbf{x}, t + |\mathbf{x}|/c_0) = K(\mathbf{x}, t)/(4\pi |\mathbf{x}|). \qquad (4.11)$$

Moreover, we emphasize that the Fourier transform of a real and even (real and odd) function is real and even (imaginary and odd). Since $\mathcal{G}$ and $\mathcal{G}_0$ are real valued, K must be real valued and consequently the real part $\mathrm{Re}(\beta^*)$ of β^* has to be even with respect to the frequency ω and $\mathrm{Im}(\beta^*)$ has to be odd with respect to ω. Attenuation is caused if $\mathrm{Re}(\beta^*)$ is positive and since then β^* has a nonzero imaginary part due to the Kramers-Kronig relation, attenuation causes dispersion. In the literature the following product ansatz is commonly used

$$\beta^*(|\mathbf{x}|, \omega) = \alpha^*(\omega) |\mathbf{x}| \qquad \omega \in \mathbb{R}, \mathbf{x} \in \mathbb{R}^3. \qquad (4.12)$$

In the sequel we concentrate on these models and use the following terminology:

Definition 4.1. *We call β^* of standard form if (4.12) holds. Then the function*

$$\alpha^* : \mathbb{R} \to \mathbb{C} \qquad (4.13)$$

is called standard attenuation coefficient *and $\alpha = \mathrm{Re}(\alpha^*)$ is called the* attenuation law. *We also call β^* the* attenuation coefficient.

From the relation (4.12), it follows that $\mathrm{Re}(\alpha^*)$ is even, $\mathrm{Im}(\alpha^*)$ is odd, and $\mathrm{Re}(\alpha^*) > 0$ (the last inequality guarantees attenuation).

In the following we summarize common attenuation coefficients and laws: In what follows α_0 denotes a positive parameter and

$$\tilde{\alpha}_0 = \frac{\alpha_0}{\cos\left(\frac{\pi}{2}\gamma\right)} \qquad (0 < \gamma \notin \mathbb{N}) \qquad (4.14)$$

is a possibly non-positive coefficient:

- **Frequency Power Laws:**
  - Let $0 < \gamma \notin \mathbb{N}$. The frequency power law *attenuation coefficient* is defined by

    $$\alpha_{pl}^*(\omega) = \tilde{\alpha}_0(-\mathrm{i}\,\omega)^\gamma = \tilde{\alpha}_0\,|\omega|^\gamma\left(\cos\left(\frac{\pi}{2}\gamma\right) - \mathrm{i}\,\mathrm{sgn}(\omega)\sin\left(\frac{\pi}{2}\gamma\right)\right) \quad (4.15)$$

    for $\omega \in \mathbb{R}$. Therefore, the *attenuation law* is given by

    $$\alpha_{pl}(\omega) = \alpha_0\,|\omega|^\gamma\ . \qquad (4.16)$$

    These models have been considered for instance in [39, 40, 43, 44].
  - Let $\gamma = 1$ and $\omega_0 \neq 0$, the attenuation coefficient is defined by

    $$\alpha_{pl}^*(\omega) := \alpha_0\,|\omega| + \mathrm{i}\,\frac{2}{\pi}\,\alpha_0\,\omega\,\log\left|\frac{\omega}{\omega_0}\right| \qquad \omega \in \mathbb{R}\,. \qquad (4.17)$$

    The attenuation law is

    $$\alpha_{pl}(\omega) := \alpha_0\,|\omega|\ . \qquad (4.18)$$

    This model has been considered in [40, 44].
- **Szabo:** Let $0 < \gamma \notin \mathbb{N}$. The attenuation coefficient[1] of Szabo's law is defined by

    $$\alpha_{sz}^*(\omega) = \frac{1}{c_0}\,\sqrt{(-\mathrm{i}\,\omega)^2 + 2\tilde{\alpha}_0 c_0(-\mathrm{i}\,\omega)^{\gamma+1}} + \mathrm{i}\,\frac{\omega}{c_0}\,. \qquad (4.19)$$

    We denote Szabo's attenuation law by

    $$\alpha_{sz}(\omega) := \mathrm{Re}(\alpha_{sz}^*(\omega))\,.$$

    For small frequencies $\alpha_{sz}(\omega)$ behaves like $\alpha_0\,|\omega|^\gamma$. This model has been considered in [39, 40] where, in addition, also a model for $\gamma \in \mathbb{N}$ has been introduced.

---

[1] In this chapter the root of a complex number is always the one with non-negative real part.

- **Thermo-Viscous Attenuation Law:** (see e.g. [21,39]): Here, for $\tau_0 > 0$, the attenuation coefficient is defined by

$$\alpha_{tv}^*(\omega) = \frac{-i\,\omega}{c_0\,\sqrt{1 - i\,\tau_0\,\omega}} + \frac{i\,\omega}{c_0} \tag{4.20}$$

  with attenuation law

$$\alpha_{tv}(\omega) = \frac{\tau_0\,\omega^2}{\sqrt{2}\,c_0\,\sqrt{\left(1 + \sqrt{1 + (\tau_0\,\omega)^2}\right)\left(1 + (\tau_0\,\omega)^2\right)}}. \tag{4.21}$$

  For small frequencies $\alpha_{tv}(\omega)$ behaves like $\frac{\tau_0\,\omega^2}{2\,c_0}$. That is the thermo-viscous law approximates a power attenuation law with exponent 2.

- **Nachman, Smith, and Waag [29]:** Consider a homogeneous and isotropic fluid with density ρ_0 in which N relaxation processes take place. Then the attenuation coefficient of the model in [29] reads as follows:

$$\alpha_{nsw}^*(\omega) = \frac{-i\,\omega}{c_0}\left[\frac{c_0}{\tilde{c}_0}\sqrt{\frac{1}{N}\sum_{m=1}^{N}\frac{1 - i\,\tilde{\tau}_m\,\omega}{1 - i\,\tau_m\,\omega}} - 1\right]. \tag{4.22}$$

  All parameters appearing in (4.22) are positive and real. κ_m and τ_m denote the compression modulus and the relaxation time of the mth relaxation process, respectively, and

$$\tilde{c}_0 := \frac{c_0}{\sqrt{1 + \sum_{m=1}^{N} c_0^2\,\rho_0\,\kappa_m}} \quad \text{and} \quad \tilde{\tau}_m := \tau_m\,(1 - N\,\tilde{c}_0^2\,\rho_0\,\kappa_m). \tag{4.23}$$

  The last two definitions imply that

$$\frac{\tilde{c}_0^2}{c_0^2} = \frac{1}{N}\sum_{m=1}^{N}\frac{\tilde{\tau}_m}{\tau_m}. \tag{4.24}$$

  We denote the according attenuation law by[2]

$$\alpha_{nsw}(\omega) := \mathrm{Re}(\alpha_{nsw}^*(\omega)).$$

- **Greenleaf and Patch [32]** consider for $\gamma \in \{1, 2\}$ the attenuation coefficient

$$\alpha_{gp}^*(\omega) = \alpha_0\,|\omega|^\gamma,$$

  which, since it is real, equals the attenuation law

$$\alpha_{gp}(\omega) = \mathrm{Re}(\alpha_{gp}^*(\omega)). \tag{4.25}$$

---

[2] In [29] they use the notion c_∞ for c_0 and c for $\tilde{c}_0$.

- **Chen and Holm [6]:** This model describes the attenuation as a function of the absolute value of the vector-valued wave number $\mathbf{k} \in \mathbb{R}^3$ (instead of the frequency $\omega \in \mathbb{R}$). Let $\mathcal{F}_{3D}$ denote the $3D$−Fourier transform

$$\mathcal{F}_{3D}\{f(\mathbf{k})\}(\mathbf{x}) = \frac{1}{\sqrt{(2\pi)^3}} \int_{\mathbb{R}^3} \exp(i\,\mathbf{x} \cdot \mathbf{k})\, f(\mathbf{k})\, d\mathbf{k} \,,$$

then the Green function of the attenuated equation is defined by

$$G(\mathbf{x}, t) = \frac{H(t)\, c_0^2}{(2\pi)^{3/2}} \mathcal{F}_{3D}\left\{ \exp(A(\cdot)\,t)\, \frac{\sin(B(\cdot)\,t)}{B(\cdot)} \right\}(\mathbf{x}) \qquad (4.26)$$

where, for given $\alpha_1 > 0$,

$$A(\mathbf{k}) := -\alpha_1\, c_0\, |\mathbf{k}|^\gamma\,, \qquad B(\mathbf{k}) := c_0 \sqrt{|\mathbf{k}|^2 - \alpha_1^2\, |\mathbf{k}|^{2\gamma}}\,. \qquad (4.27)$$

- **In [22]** we proposed

$$\alpha_{ksb}^*(\omega) = \frac{\alpha_0\,(-i\,\omega)}{c_0 \sqrt{1 + (-i\,\tau_0\,\omega)^{\gamma-1}}} \qquad (\gamma \in (1,2],\ \tau_0 > 0)\,, \qquad (4.28)$$

where the square root is again the complex root with positive real part. Let $\gamma \in (1,2]$. Then, for small frequencies we have

$$\alpha_{ksb}(\omega) \approx \frac{\alpha_0\, \sin(\frac{\pi}{2}(\gamma-1))}{2\, c_0\, \tau_0}\, |\tau_0\,\omega|^\gamma > 0\,.$$

Thus our model behaves like a power law for small frequencies.

Distinctive features of unattenuated wave propagation (i.e. the solution of the standard wave equation) are *causality* and *finite wave front velocity*. It is reasonable to assume that the attenuated wave satisfies the same distinctive properties as well. In the following we analyze causality properties of the standard attenuation models.

4.3 Causality

In the following we present some abstract definitions and basic notations. In the remainder $\mathbf{x}$ will always denote a vector in three dimensional space. When we speak about functions, we always mean generalized functions, such as for instance distributions or tempered distributions – we recall the definitions of (tempered) distribution in the course of the chapter.

Definition 4.2. *A function $f := f(\mathbf{x}, t)$ defined on the Euclidean space over time (i.e. in $\mathbb{R}^4$) is said to be* causal *if it satisfies $f(\mathbf{x}, t) = 0$ for $t < 0$.*

Notation & Terminology 4.1 *Let $A : D \to D$ be a linear operator, where $\emptyset \neq D$ is an appropriate set of functions from $\mathbb{R}^4$ to $\mathbb{R}$. In this chapter we always assume that A satisfies the following properties:*

- *A is shift invariant in space and time. That is, for every function f and every shift $L := L(\mathbf{x}, t) := (\mathbf{x} - \mathbf{x}_0, t - t_0)$, with $\mathbf{x}_0 \in \mathbb{R}^3$ and $t_0 \in \mathbb{R}$, it holds that*

$$A(f \circ L) = (Af) \circ L .$$

- *A is rotation invariant in space. That is, for every function f and every rotation matrix R, it holds that*

$$A(Rf) = R(Af) .$$

- *A is causal. That is, it maps causal functions to causal functions. From (4.29) it follows that A is causal, if and only if the associated Green function is causal.*

Definition 4.3. *The Green function of A is defined by*

$$\mathcal{G} := \mathcal{G}(\mathbf{x}, t) = A\delta_{\mathbf{x},t}(\mathbf{x}, t) .$$

Remark 4.1. The operator A is uniquely determined by $\mathcal{G}$ and vice versa. This follows from the fact that

$$
\begin{aligned}
Af(\mathbf{x}_0, t_0) &= A \left(\int_{\mathbb{R}} \int_{\mathbb{R}^3} f(\mathbf{x}_0 - \mathbf{x}, t - t_0)\delta_{\mathbf{x},t}(\mathbf{x}, t) \, d\mathbf{x}dt \right) \\
&= \int_{\mathbb{R}} \int_{\mathbb{R}^3} f(\mathbf{x}_0 - \mathbf{x}, t - t_0)\mathcal{G}(\mathbf{x}, t) \, d\mathbf{x}dt .
\end{aligned}
\tag{4.29}
$$

Moreover, we use the following terminology and abbreviations:

- *From the rotation invariance of A it follows that*

$$\hat{T}(\mathbf{x}) := \sup\{t : \mathcal{G}(\mathbf{x}, \tau) = 0 \text{ for all } \tau \leq t\} , \tag{4.30}$$

is rotationally symmetric, which allows us to use the shorthand notation

$$T(r) = \hat{T}(\mathbf{x}) \text{ where } r = |\mathbf{x}| . \tag{4.31}$$

With this notation (4.30) can be equivalently expressed as

$$\mathcal{G}(\mathbf{x}, t + T(|\mathbf{x}|)) = 0 \text{ for every } t < 0 . \tag{4.32}$$

In physical terms $T(|\mathbf{x}|)$ denotes the travel time of a wave front originating at position $\mathbf{0}$ at time $t = 0$ and traveling to $\mathbf{x}$.

- *Because $\mathcal{G}$ is rotationally symmetric we can write*

$$\mathcal{G}(\mathbf{x}, T(|\mathbf{x}|)) = \hat{\mathcal{G}}(r, T(r)) \ \text{with} \ r = |\mathbf{x}| \ .$$

Taking the inverse function of T, which we denote by $S = S(t)$, we then find

$$\mathcal{G}(\mathbf{x}, T(|\mathbf{x}|)) = \hat{\mathcal{G}}(S(t), t) \,,$$

- *The wave front is the set*

$$\mathcal{W} := \{(\mathbf{x}, T(|\mathbf{x}|)) : \mathbf{x} \in \mathbb{R}^3\}$$

- *The wave front speed V is the variation of the location of the wave front as a function of time. That is,*

$$V(t) = \frac{dS}{dt}(t) = \frac{1}{T'(r)}\bigg|_{r=S(t)} . \tag{4.33}$$

Here T' denotes the derivative with respect to the radial component r.
- *We say that $\mathcal{A}$ has a finite speed of propagation if there exists a constant $\hat{c}_0$ such that*

$$0 < (T'(r))^{-1} \le \hat{c}_0 < \infty \ . \tag{4.34}$$

In this case it follows from (4.33) that the wave front velocity satisfies

$$V(t) \le \hat{c}_0 < \infty \ . \tag{4.35}$$

- *We call an operator $\mathcal{A}$ strongly causal, if it is causal and satisfies the finite propagation speed property.*

The following lemma addresses the case of attenuation coefficients of standard form and gives examples of strongly causal operators $\mathcal{A}$.

Lemma 4.1. *Let $\beta^*(|\mathbf{x}|, \omega) = \alpha^*(\omega)|\mathbf{x}|$ be of the standard form (4.12) and $\mathcal{A}$ (4.29) be the operator defined by the Green function $\mathcal{G}$, which is defined in (4.9). Then $\mathcal{A}$ is strongly causal if and only if for every $\mathbf{x} \in \mathbb{R}^3$ the function*

$$t \to \frac{1}{\sqrt{2\pi}} \mathcal{F}^{-1} \{\exp(-\alpha^*(\omega)|\mathbf{x}|)\} \,,$$

defined in (4.10), is causal.

Proof. We assume that $\mathcal{A}$ is strongly causal. It follows from [22, Theorem 3.1] that there exists a constant c, which is smaller than or equal to the wave speed c_0 from (4.4), which satisfies $T(|\mathbf{x}|) = \frac{|\mathbf{x}|}{c}$ for all $\mathbf{x} \in \mathbb{R}^3$. Using the definitions of the travel time $T(|x|)$ and (4.10), it follows from (4.11) that $t \to K(\mathbf{x}, t)$ is causal.

Now, for every $\mathbf{x} \in \mathbb{R}^3$ let K be causal. Then from (4.11) it follows that $t \to \mathcal{G}(\mathbf{x}, t + |\mathbf{x}|/c_0))$ is causal. Since $T(|\mathbf{x}|)$ denotes the largest positive time period for which $t \to \mathcal{G}(\mathbf{x}, t + T(|\mathbf{x}|))$ is causal, we have for $r > 0$:

$$0 < \frac{r}{c_0} \le T(r) = \int_0^r \frac{1}{V(s)} \, . \qquad (4.36)$$

Here V is parameterized with respect to the distance s at time t of the wave front from its origin. As shown in the proof of [22, Theorem 3.1], the fact that β^* is of standard form together with (4.36) implies that there exist a constant c such that $T(r) = r/c$ for all $r > 0$. But then from (4.36) it follows $0 < r/c_0 \le r/c < \infty$ and consequently $0 < c \le c_0 < \infty$.

Finally we explain the above notation for the standard wave equation:

Remark 4.2. In the case of the standard wave equation the wave front is the support of the Green function $\mathcal{G}_0$, the wave front velocity is c_0, and $T(|\mathbf{x}|) = \frac{|\mathbf{x}|}{c_0}$ denotes the travel time of the wave front.

4.4 Strong Causality of Attenuation Laws

In this section we analyze causality properties of attenuation laws. We split the section into two parts, where the first concerns numerical studies to determine the kernel function K, defined in (4.10), and the second part contains analytical investigations.

In Figs. 4.1, 4.2 and 4.3 we represent the attenuation kernels according to power, Szabo's, and the thermo-viscous law.

The figures already indicate that power laws with index greater than 1 violate causality. In the following we support these computational studies by analytical considerations. Thereby we make use of distribution theory, which we recall first. Generally speaking *Distributions* are generalized functions:

Definition 4.4. We use the abbreviations:

- $\mathcal{D} := C_0^\infty(\mathbb{R}, \mathbb{C})$ is the space of infinitely often differentiable functions from $\mathbb{R}$ to $\mathbb{C}$ which have compact support.
- $\mathcal{S}$ is the space of infinitely often differentiable functions from $\mathbb{R}$ to $\mathbb{C}$ which are *rapidly decreasing*. A function $f : \mathbb{R} \to \mathbb{C}$ is rapidly decreasing if for all $i, j \in \mathbb{N}_0$

$$|x|^i \left| f^{(j)}(x) \right| \to 0 \text{ for } |x| \to \infty \, .$$

- $\mathcal{S}$ is a locally convex space (see [54] for a definition) with the topology induced by the family of semi-norms

$$p_{P,j}(f) = \sup_{x \in \mathbb{R}} \left| P(x) f^{(j)}(x) \right| ,$$

where P is a polynomial and $j \in \mathbb{N}_0$. The topology on a locally convex set is defined as follows: $U \subseteq \mathcal{S}$ is open, if for every $f \in U$ there exists $\varepsilon > 0$

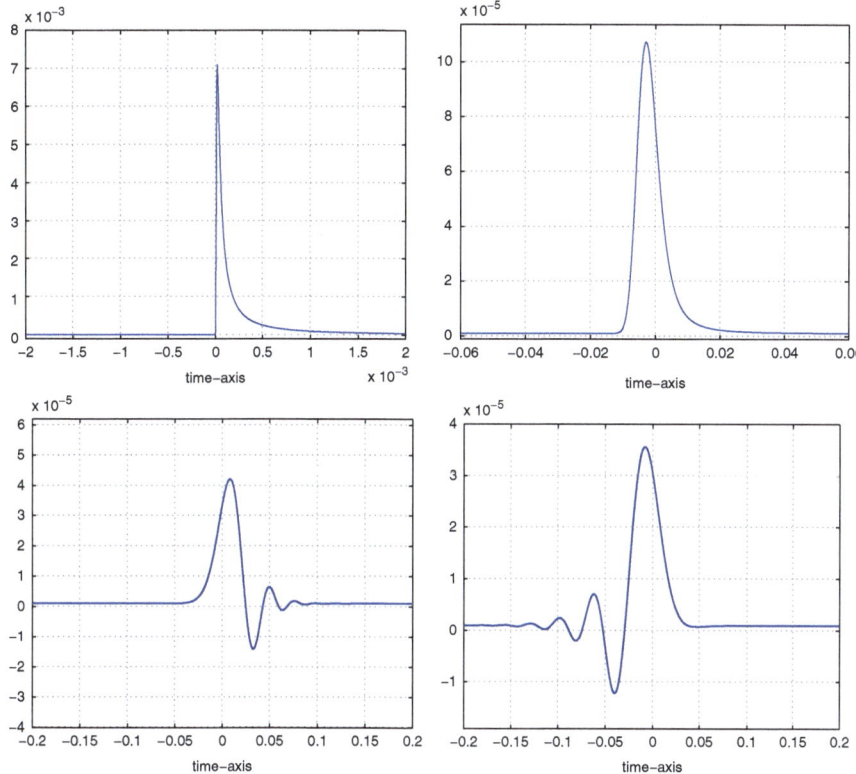

Fig. 4.1. Simulation of $K(\mathbf{x}, t)$ for the frequency power law with $(\gamma, \alpha_0) \in \{(0.5, 0.1581), (1.5, 0.0316), (2.7, 0.0071), (3.3, 0.0027)\}$, $c_0 = 1$ and $|\mathbf{x}| = \frac{1}{4}$. In the first example $\gamma < 1$ and thus the function is causal. For all other cases it is non causal

and a finite non-empty set J' of polynomials and a finite set of indices K' such that

$$\bigcap_{P \in J', k \in K'} \{g \in \mathcal{S} : p_{P,k}(g - f) < \varepsilon\} \subseteq U .$$

- The space of *tempered distributions*, $\mathcal{S}'$, is the space of linear continuous functionals and $\mathcal{S}$.
- A functional $L : \mathcal{S} \to \mathbb{C}$ is continuous if there exists a constant $C > 0$ and a seminorm $p_{P,j}$ such that

$$|Lu| \leq C p_{P,j}(u) , \text{ for every } u \in \mathcal{S}$$

(see [54, Sect. I.6, Thm. 1]

In the following we give some examples of tempered distributions and review some of their properties. The examples are taken from [54, Sec6.2, Ex. 3] and [9, Remark 6].

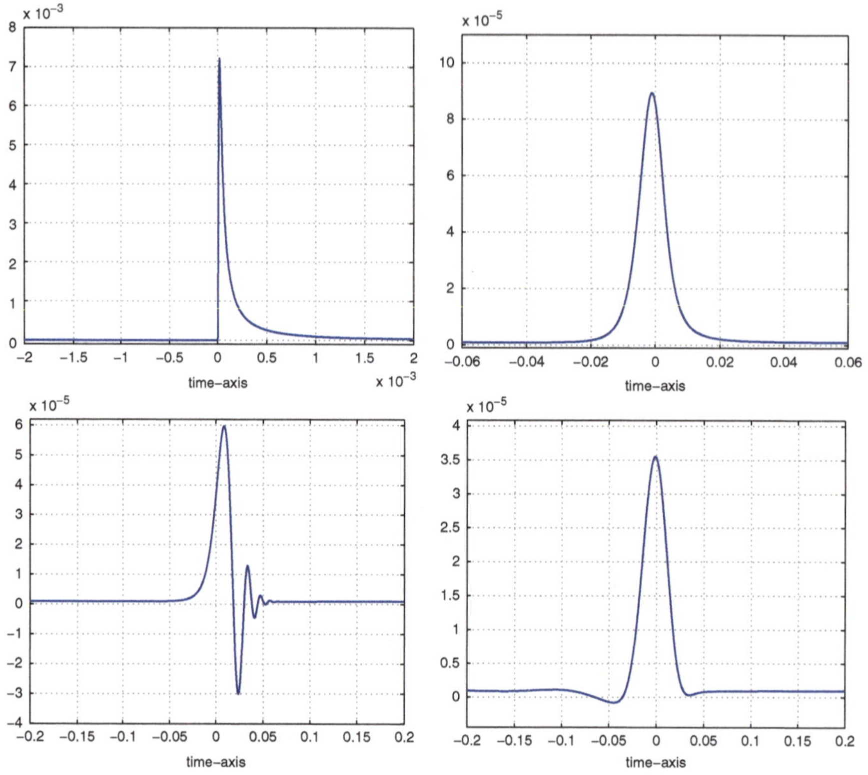

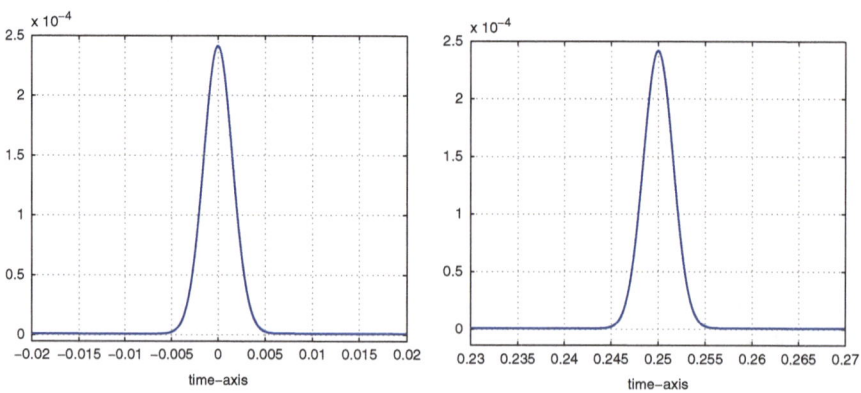

Fig. 4.2. Simulation of $K(\mathbf{x},t)$ for Szabo's frequency law with $(\gamma,\alpha_0) \in \{(0.5, 0.1581), (1.5, 0.0316), (2.7, 0.0071), (3.3, 0.0027)\}$, $c_0 = 1$ and $|\mathbf{x}| = \frac{1}{4}$

Fig. 4.3. *Left:* $K(\mathbf{x},t)$ defined by the complex thermo-viscous attenuation law with $\tau_0 = 10^{-5}$, $c_0 = 1$ and fixed $|\mathbf{x}| = \frac{1}{4}$. *Right:* The proposed law (4.28) for $\gamma = 2$ with $\alpha_1 = 1$, $\tau_0 = 10^{-5}$, $c_0 = 1$ and fixed $|\mathbf{x}| = \frac{1}{4}$ is causal

Example 1. (**Examples of Tempered Distributions**)

- Let $1 \leq p \leq \infty$ and $f \in L^p(\mathbb{R}, \mathbb{C})$, then the linear operator $T\phi = \int_{\mathbb{R}} f(x)\phi(x)\,dx$ is a tempered distribution. In the following we identify f and T, and this clarifies the terminology $f \in \mathcal{S}'$ later on.
- $\mathcal{S} \subseteq \mathcal{S}'$ – thereby already the above relation between functions and tempered distributions is used.
- Distributions with compact support are tempered distributions. For instance the δ-Distribution is a tempered distribution.
- Polynomials are tempered distributions.
- The functions f of $L^1_{loc}(\mathbb{R})$ which are uniformly bounded by a polynomial for $|x|$ sufficiently large, are tempered distributions.[3]

Lemma 4.2. • *The pointwise limit $f : \mathbb{R} \to \mathbb{C}$ of a sequence of functions $\{f_n : \mathbb{R} \to \mathbb{C}\} \subseteq \mathcal{S}'$, is again a tempered distribution.*
- *Let $f \in \mathcal{S}'$, then $\mathcal{F}\{f\} \in \mathcal{S}'$ and $\mathcal{F}^{-1}\{f\} \in \mathcal{S}'$.*

In the following we review Theorem 4 on p294 ff from [10] which characterized when a generalized function $f \in \mathcal{S}'(\mathbb{R})$ is causal, that is, when $\operatorname{supp}(f) \subseteq [0, \infty)$. Below we use the following notation

$$\mathbb{C}_\varepsilon := \{z \in \mathbb{C} : \operatorname{Im}(z) \geq \varepsilon\}\,.$$

Theorem 4.2. *(Theorem 4 on p294 ff in [10]) Let $f \in \mathcal{S}'(\mathbb{R})$. Then f is causal[4] if and only if*

1. *There exists a function $F : \mathbb{C}_0 := \{\xi + i\eta : \eta \geq 0\} \to \mathbb{C}$, which is holomorphic in the interior $\mathring{\mathbb{C}}_0 := \{\xi + i\eta : \eta > 0\}$.[5]*
2. *For all fixed $\eta > 0$ and $\xi \in \mathbb{R}$, $F(\xi + i\eta)$ is a tempered distribution with respect to the variable ξ and for $\eta \to 0$ $F(\xi + i\eta)$ is convergent (with respect to the weak topology on $\mathcal{S}'$) to $\mathcal{F}\{f\} : \mathbb{R} \to \mathbb{C}$.[6]*
3. *For every $\varepsilon > 0$, there exists a polynomial P such that*

$$|F(z)| \leq P(|z|) \qquad for \qquad z \in \mathbb{C}_\varepsilon\,.$$

Remark 4.3. The definition of the Fourier transform in this chapter has a different sign as in [10] and consequently also $\mathbb{C}_0$ denotes the upper half plane and not the lower half plane as in [10].

For analyzing attenuation laws, we use the following corollary, which is derived from Theorem 4.2.

---

[3] A function f is an element of $L^1_{loc}(\mathbb{R})$ if it is in L^1 on every compact set.

[4] In Theorem 4 on p294 ff in [10] the assumption that f is strongly causal is expressed by $f \in \mathcal{D}'_+$, which is the set of distributions with support in $[0, +\infty)$.

[5] A function $F : \mathbb{C}_0 \to \mathbb{C}$ is holomorphic in $\mathring{\mathbb{C}}_0$ if it is complex differentiable in $\mathring{\mathbb{C}}_0$. Sometimes the functions are also refered to as analytic or regular functions or conformal maps.

[6] A function $F : \mathbb{C}_0 \to \mathbb{C}$ which satisfies Items 1,2 of Theorem 4.2 is called *holomorphic extension* of $\mathcal{F}\{f\}$.

Corollary 4.1. *Let $\alpha^* : \mathbb{R} \to \mathbb{C}$ be continuous and let there exist a holomorphic extension to $\mathbb{C}_0$, which for the sake of simplicity of notation is again denoted be α^*. In addition, let $\alpha^*(\xi + i\eta) \to \alpha^*(\xi)$ for $\eta \to 0$ pointwise. We denote by $\alpha : \mathbb{C}_0 \to \mathbb{C}$ the real part of α^*.[7] Moreover, we assume that there exists a constant C such that*

$$\alpha(\omega) \geq C \text{ for all } \omega \in \mathbb{R}. \tag{4.37}$$

1. If in addition

$$\alpha(z) \geq C \text{ for all } z \in \mathbb{C}_0. \tag{4.38}$$

Then, for every $\mathbf{x} \in \mathbb{R}^3$, the function

$$t \to K(\mathbf{x}, t) := \frac{1}{\sqrt{2\pi}} \mathcal{F}^{-1} \left\{ \exp\left(-\alpha^*(\cdot) \, |\mathbf{x}|\right) \right\}(t) \tag{4.39}$$

is causal.

2. On the other hand, if there exists $C_1 > 0$, $\mu > 0$ and $C_2 \in \mathbb{R}$ and a sequence $\{z_n\}$ in $\overset{\circ}{\mathbb{C}}_0$ such that

$$\alpha(z_n) \leq -C_1 \, |z_n|^\mu - C_2, \tag{4.40}$$

then K violates causality.

Proof. Let $\mathbf{x} \in \mathbb{R}^3$ fixed. We apply Theorem 4.2 to $f(\cdot) = K(\mathbf{x}, \cdot)$. Therefore, we have

$$\sqrt{2\pi} \mathcal{F}\{f\}(\omega) = \exp\left(-\alpha^*(\omega) \, |\mathbf{x}|\right).$$

Under the assumption (4.37), taking into account that α_{pl}^* is continuous, $\mathcal{F}\{f\}$ is in $L_{loc}^1(\mathbb{R})$ and bounded by a constant polynomial, thus in $\mathcal{S}'$ (cf. Example 4.2), and consequently, according to Lemma 4.2, $f \in \mathcal{S}'$. Therefore, the general assumption of Theorem 4.2 is satisfied.

The function $z \in \mathbb{C}_0 \mapsto F(z) := \exp\left(-\alpha^*(z) \, |\mathbf{x}|\right)$ is an extension of $2\pi \mathcal{F}\{f\}(\omega)$, which is holomorphic in $\overset{\circ}{\mathbb{C}}_0$. Thus Item 1 of Theorem 4.2 holds.

- For proving the first assertion, it follows from (4.38) that for $z = \xi + i\eta$ with $\eta \geq 0$,

$$|F(z)| \leq \exp\left(-C \, |\mathbf{x}|\right), \tag{4.41}$$

which, in particular, shows that for all $\eta \geq 0$, the functions $\xi \to F(\xi + i\eta)$ is a tempered distribution (cf. Lemma 4.2). Hence Item 1 of Theorem 4.2 holds.

Moreover, since by assumption $\alpha^*(\xi + i\eta) \to \alpha^*(\xi)$ for $\eta \to 0$ pointwise, $F(\xi + i\eta)$ converges to $F(\xi) = 2\pi \mathcal{F}\{f\}(\xi)$ pointwise. Because the limit is a tempered distribution and the convergence is with respect to the weak topology $\mathcal{S}'$ (cf. Lemma 4.2). Hence Item 2 of Theorem 4.2 holds.

---

[7] $\alpha : \mathbb{C}_0 \to \mathbb{C}$ extends the function $\omega \in \mathbb{R} \to \alpha(\omega)$ but is not an holomorphic extension.

Moreover, from (4.41) it follows that $|F(z)|$ is bounded by a constant polynomial. Hence Item (3) of Theorem 4.2 holds and therefore Theorem 4.2 guarantees that $t \mapsto K(\mathbf{x}, t)$ is causal.

- For the second case, Item 3 of Theorem 4.2 is violated. Consequently, $K(\mathbf{x}, \cdot)$ is not causal.

Power Laws

Theorem 4.3. *Let $0 < \gamma \in \mathbb{R}$, be not an odd number, and $\omega \in \mathbb{R} \mapsto \alpha_{pl}^*(\omega) = \tilde{\alpha}_0(-i\,\omega)^\gamma$ be the power law attenuation coefficient from (4.15), with $\tilde{\alpha}_0 = \alpha_0/\cos\left(\frac{\pi}{2}\gamma\right)$ as in (4.14). Then, the function K, defined in (4.39), is causal if and only if $\gamma \in (0, 1)$.*

Proof. Let $\mathbf{x} \in \mathbb{R}^3$ be fixed. The function $z \in \mathbb{C} \mapsto \alpha_{pl}^*(z) = \tilde{\alpha}_0(-i\,z)^\gamma$ is the holomorphic extension of $\omega \in \mathbb{R} \mapsto \alpha_{pl}^*(\omega)$. We prove or disprove causality by using Corollary 4.1.

For $z = |z| \exp(i\,\phi)$ it follows from (4.15) that

$$\mathrm{Re}((-i\,z)^\gamma) = \mathrm{Re}(|z|^\gamma \exp(i\,\gamma\,(\phi - \pi/2))) = |z|^\gamma \cos(\gamma(\phi - \pi/2)) \,.$$

This implies that

$$\alpha_{pl}(z) = \tilde{\alpha}_0\mathrm{Re}((-i\,z)^\gamma) = \tilde{\alpha}_0 \cos(\gamma(\phi - \pi/2)) |z|^\gamma \,. \tag{4.42}$$

In particular, if $z = \omega \in \mathbb{R}$, then ϕ is either 0 or π. Taking into account the definition of $\tilde{\alpha}_0$ and that the cos-function is symmetric around the origin, it follows that

$$\alpha_{pl}(\omega) = \alpha_0 |w|^\gamma \geq 0 \,. \tag{4.43}$$

Thus (4.37) holds.

- Let $\gamma \in (0, 1)$: Every $z = |z| \exp(i\,\phi) \in \mathbb{C}_0$ satisfies $\phi \in [0, \pi]$. Consequently $\gamma(\phi - \pi/2) \in [-\pi/2, \pi/2]$ and thus $\cos(\gamma(\phi - \pi/2))$ is uniformly non-negative. Even more for $\gamma \in (0, 1)$ the coefficient $\tilde{\alpha}_0$, defined in (4.14), is positive. In summary, we have that there exists a constant $C_1 \geq 0$ such that

$$\alpha_{pl}(z) \geq C_1 |z|^\gamma \geq 0 \qquad \text{for} \qquad z \in \mathbb{C}_0 \,. \tag{4.44}$$

Thus (4.38) holds and application of Corollary 4.1 shows that K is causal.
- Let $\gamma \in (1, 3) \cup (5, 7) \cup \cdots$. Then $\tilde{\alpha}_0 < 0$. The sequence

$$\{z_n := n \exp(i\,\pi/2) = i\,n\}_{n \in \mathbb{N}} \tag{4.45}$$

consists of elements of $\mathring{\mathbb{C}}_0$ and satisfies assumption (4.40), that is,

$$\alpha_{pl}(z_n) = \underbrace{\tilde{\alpha}_0}_{<0} |z_n|^\gamma \,. \tag{4.46}$$

Application of Corollary 4.1 shows that K is not causal.

- Let $\gamma \in (3,5) \cup (7,9) \cup \cdots$. We fix some $0 < \delta < \pi/2$, and define

$$\phi := \left(1 + \frac{1}{\gamma}\right)\frac{\pi}{2} + \frac{\delta}{\gamma}.$$

The sequence

$$\{z_n := n \exp(i\phi)\} \tag{4.47}$$

consists of elements of $\mathring{\mathbb{C}}_0$. Under the above assumptions, it follows that $\tilde{\alpha}_0 > 0$ and therefore

$$\alpha_{pl}(z_n) = \underbrace{\tilde{\alpha}_0}_{>0} \underbrace{\cos(\pi/2 + \delta)}_{<0} |z_n|^\gamma . \tag{4.48}$$

Thus form Corollary 4.1 the assertion follows.

In the following we analyze the following family of variants of power laws:

$$\alpha_{pl+}^*(\omega) = \tilde{\alpha}_0(-i\omega)^\gamma + \alpha_1(-i\omega), \tag{4.49}$$

which have been considered in [40, 43].

Theorem 4.4. *Let* $0 < \gamma \notin \mathbb{N}$ *and* α_{pl+}^* *as defined in (4.49). Moreover, let* K *be as in (4.39). Then, if* $\gamma > 1$, K *is not causal. For* $\gamma \in (0,1)$ K *is causal if and only if* $\alpha_1 \in [0,\infty)$.

Proof. The holomorphic extension of $\omega \in \mathbb{R} \to \alpha_{pl+}^*(\omega)$ is the function

$$\alpha_{pl+}^*(z) = \tilde{\alpha}_0(-iz)^\gamma + \alpha_1(-iz)$$

and consequently

$$\alpha_{pl+}^*(z) = \tilde{\alpha}_0 |z|^\gamma \left(1 + \frac{\alpha_1}{\tilde{\alpha}_0}|z|^{1-\gamma}\right)\cos\left(\gamma\left(\phi - \frac{\pi}{2}\right)\right).$$

- For $\gamma > 1$ we have that $1 + \frac{\alpha_1}{\tilde{\alpha}_0}|z|^{1-\gamma} \to 1$ for $|z| \to \infty$. Let $\{z_n\}$ as defined in (4.45) or (4.47). Then, since for both sequences $|z_n| \to \infty$, it follows from (4.46), (4.48) that the according sequences $\{z_n\}$ satisfy (4.40) for n sufficiently large, respectively. Thus K is not causal.
- For $\gamma \in (0,1)$ and $\alpha_1 \geq 0$ the assertion follows already from the fact that $\alpha_{pl+}(\omega) \geq \alpha_{pl}(\omega)$ and that the later already satisfies (4.38). Thus K is causal.
- Let $\gamma \in (0,1)$ and $\alpha_1 < 0$. Then for some $0 < \delta < -\alpha_1$ fixed, we can find a constant C_2 such that for all $z \in \mathbb{C}$

$$\tilde{\alpha}_0 |z|^\gamma + \alpha_1 |z| \leq \underbrace{(\alpha_1 + \delta)}_{<0} |z| - C_2 .$$

Consequently, for $\{z_n = in\}$, we have

$$\alpha_{pl+}(z_n) \leq (\alpha_1 + \delta)|n| - C_2 .$$

which shows (4.40). Thus K is not causal.

Powerlaw with $\gamma = 1$

Theorem 4.5. Let α_{pl}^* be as in defined in (4.17). Then the function K, defined in (4.39) is not causal.

Proof. First, we prove that

$$z \in \mathbb{C}_0 \mapsto \hat{\alpha}_{pl}^*(z) := \alpha_0 z + i \frac{2\alpha_0}{\pi} z \log\left(\frac{z}{w_0}\right)$$

is the holomorphic extension of $w \to \alpha_{pl}^*(w)$. This assertion follows from the facts

$$\hat{\alpha}_{pl}^*(w) = \alpha_{pl}^*(w) \text{ for } \qquad w > 0,$$

$$\lim_{\eta \to 0+} i \frac{2}{\pi} \log\left(\frac{w + i\eta}{w_0}\right) = i \frac{2}{\pi} \log\left|\frac{w}{w_0}\right| - 2 \text{ for } w < 0.$$

Since

$$\alpha_{pl}(w) = \mathrm{Re}(\alpha_{pl}^*(w)) = \alpha_0 |w| \geq 0,$$

Corollary 4.1 is applicable. For the elements of the sequence $\{z_n := in\}_{n \in \mathbb{N}}$ in $\mathbb{C}_0$

$$\hat{\alpha}_{pl}^*(z_n) = -\frac{2\alpha_0}{\pi} n \log\left(\frac{n}{w_0}\right).$$

is real and therefore equals $\alpha_{pl}(z_n)$ and thus (4.40) holds. Thus Corollary 4.1 gives the assertion.

Szabo's Model:

Proposition 4.1. For $\alpha_0 > 0$ let α_{sz}^* be the coefficient of Szabo's model (4.19). Then, for $\gamma \in (0, 1)$, the function K (4.10) is a causal function and for $\gamma > 1$ with $\gamma \notin \mathbb{N}$, K violates causality.

Proof. Without loss of generality we assume that $c_0 = 1$. The holomorphic extension of $\alpha_{sz}^* : \mathbb{R} \to \mathbb{C}$ from (4.19) is

$$z \in \mathbb{C}_0 \to \alpha_{sz}^*(z) = (-iz)\left[\sqrt{1 + 2\tilde{\alpha}_0(-iz)^{\gamma-1}} - 1\right].$$

First, we make some general manipulations which can be used in several ways: Let $z = \xi + i\eta \in \mathbb{C}_0$. We use the polar representation

$$z = |z| \exp(i\phi), \quad \phi := \phi(z) \in [0, \pi].$$

Then

$$\alpha_{sz}^*(z) = |z| \exp(i(\phi - \pi/2)) \Psi(z),$$

where

$$\Psi(z) := \sqrt{\hat{\Psi}(z)} - 1 \,,$$
$$\hat{\Psi}(z) := 1 + 2\tilde{\alpha}_0 |z|^{\gamma-1} \exp\left(i\,\delta\right) \,,$$
$$\delta := \delta(z) := (\phi - \pi/2)(\gamma - 1) \,. \tag{4.50}$$

With this notation we have

$$\mathrm{Re}(\hat{\Psi}(z)) = 1 + 2\tilde{\alpha}_0 \cos(\delta) |z|^{\gamma-1} \,,$$
$$\mathrm{Im}(\hat{\Psi}(z)) = 2\tilde{\alpha}_0 \sin(\delta) |z|^{\gamma-1} \,,$$
$$\left|\hat{\Psi}(z)\right| = |z|^{\gamma-1} \sqrt{(1 + 2\tilde{\alpha}_0 \cos(\delta))^2 + 4\tilde{\alpha}_0^2 \sin(\delta)^2} \,. \tag{4.51}$$

Representing $\hat{\Psi}$ in polar coordinates,

$$\hat{\Psi}(z) = \left|\hat{\Psi}(z)\right| \exp\left(i\,\theta(z)\right) \,,$$

we get

$$\sqrt{\hat{\Psi}(z)} = \sqrt{\left|\hat{\Psi}(z)\right|} \exp\left(i\,\theta(z)/2\right) \qquad \text{with}$$
$$\theta(z) = \arctan(\mathrm{Im}(\hat{\Psi}(z))/\mathrm{Re}(\hat{\Psi}(z)))) \qquad \theta \in (-\pi, \pi) \,. \tag{4.52}$$

Note, that $\sqrt{\hat{\Psi}(z)}$ is the complex root with non-negative real part, which meets the general assumption of the chapter. Moreover, we have

$$\alpha_{sz}(z) = \mathrm{Re}(\alpha_{sz}^*(z)) = \eta \mathrm{Re}(\Psi(z)) + \xi \mathrm{Im}(\Psi(z)) \,. \tag{4.53}$$

First, we prove that $\mathrm{Re}(\Psi(z)) \geq 0$: We use the elementary inequality

$$\cos(\theta(z)) \leq \cos^2(\theta(z)/2) \,,$$

and $\cos(\theta(z)/2) \geq 0$ which imply that

$$\mathrm{Re}\left(\sqrt{\hat{\Psi}(z)}\right) = \sqrt{\left|\hat{\Psi}(z)\right|} \cos(\theta(z)/2)$$
$$= \sqrt{\left|\hat{\Psi}(z)\right| \cos^2(\theta(z)/2)}$$
$$\geq \sqrt{\mathrm{Re}(\hat{\Psi}(z))} \,. \tag{4.54}$$

- Now, let $\gamma \in (0, 1)$. Since $\mathrm{Re}(\hat{\Psi}(z)) \geq 1$ for $\gamma \in (0, 1)$, it follows that for all $z \in \mathbb{C}_0$

$$\eta \mathrm{Re}(\Psi(z)) = \eta \mathrm{Re}\left(\sqrt{\hat{\Psi}(z)}\right) - \eta \geq \eta \sqrt{\mathrm{Re}(\hat{\Psi}(z))} - \eta \geq 0 \,.$$

Thus $\eta \mathrm{Re}(\Psi(z)) \geq 0$.

Now we show that $z \to \xi \mathrm{Im}(\Psi(z))$ is uniformly bounded from below by 0 in $\mathbb{C}_0$. Thus according to (4.53) α_{sz} is uniformly bounded from below, and thus from Corollary 4.1, it follows that $t \to K(\mathbf{x}, t)$ is causal.

Using the definition of θ, (4.52), and the facts that $\delta \in [0, (1 - \gamma)\pi/2]$ for $\phi \in [0, \pi/2]$ and $\delta \in [(\gamma - 1)\pi/2, 0)$ for $\phi \in (\pi/2, \pi]$ it follows from the monotonicity of $\tan$ on $(-\pi, \pi)$ that

$$\theta(z) = \arctan \left(\frac{2\tilde{\alpha}_0 \sin(\delta) |z|^{\gamma-1}}{1 + 2\tilde{\alpha}_0 \cos(\delta) |z|^{\gamma-1}} \right) \in \left\{ \begin{array}{ll} [0, \delta] & \text{for all } \phi \in [0, \pi/2], \\ [\delta, 0) & \text{for all } \phi \in (\pi/2, \pi] \end{array} \right.$$

Now, noting that $\mathrm{sgn}(\xi) = \mathrm{sgn}\left(\sin(\theta(z)/2)\right)$ it follows that

$$\xi \mathrm{Im}(\Psi(z)) = \xi \sqrt{\left|\hat{\Psi}(z)\right|} \sin(\theta(z)/2) \geq 0. \tag{4.55}$$

Thus the assertion follows from Corollary 4.1.

• Assume that $\gamma > 1$. Let $z = \xi + i\eta$ with $\eta = 0$. Since the square root in (4.50) is such that $\mathrm{Re}(\Psi(z)) > 0$ for $z \in \mathbb{C}_0$, property (4.102) in the Appendix implies $\xi \mathrm{Im}(\Psi(z)) > 0$ for $z = \xi$ and hence $\alpha_{sz}(z = \xi) \geq 0$. Thus (4.37) holds and we can apply Corollary 4.1.

  – Let $\gamma \in (1, 3) \cup (5, 7) \cup \cdots$, which implies that $\cos(\gamma \pi/2) < 0$, and consequently, $\tilde{\alpha}_0 < 0$. For sufficiently large n the elements of the sequence $\{z_n := in\}$ satisfy

  $$\alpha_{sz}(z_n) = n \mathrm{Re}\left(\sqrt{1 - 2|\tilde{\alpha}_0| n^{\gamma-1}} - 1\right) \leq -n,$$

  which shows that (4.40) holds with $\mu = 1$, $C_1 = 1/2$ and $C_2 = 0$, and hence the assertion follows from Corollary 4.1.

  – Let $\gamma \in (3, 5) \cup (7, 9) \cup \cdots$, which implies that $\cos(\gamma \pi/2) > 0$, and consequently, $\tilde{\alpha}_0 > 0$. Now, let $z_n := n \exp(i\phi)$ with

  $$\phi := \frac{\pi}{\gamma - 1} + \frac{\pi}{2}.$$

  Since $\gamma > 3$, we have $\phi \in (\pi/2, \pi)$, and therefor $\mathrm{Re}(z_n) = n \cos(\phi) < 0$ and $\mathrm{Im}(z_n) = n \sin(\phi) > 0$. Moreover,

  $$\hat{\Psi}(z_n) = 1 - 2|\tilde{\alpha}_0| n^{\gamma-1}$$

  and thus for sufficiently large n we have $\mathrm{Re}(\Psi(z_n)) = -1$ and $\mathrm{Im}\left(\sqrt{\hat{\Psi}(z_n)}\right) \geq C_1 n^{(\gamma-1)/2}$ for some constant $C_1 > 0$. Hence it follows that

  $$\alpha_{sz}(z_n) = -\mathrm{Im}(z_n) + \mathrm{Re}(z_n) \mathrm{Im}(\sqrt{\hat{\Psi}(z_n)})$$
  $$= -n |\sin(\phi)| - n |\cos(\phi)| C_1 n^{(\gamma-1)/2}$$

  and therefore (4.40) holds. Thus from Corollary 4.1 the assertion follows.

Thermo-Viscous Attenuation Law

Theorem 4.6. *Let $c_0, \tau_0 > 0$ and let α_{tv}^* as defined in (4.20). Then the kernel function K violates causality.*

Proof. Since the function $z \in \mathbb{C}_0 \to 1 - i\,\tau_0\,z$ does not vanish, the function

$$z \in \mathbb{C}_0 \to \alpha_{tv}^*(z) = \frac{-i\,z}{c_0\,\sqrt{1 - i\,\tau_0 z}} + \frac{i\,z}{c_0}$$

is the holomorphic extension of $\omega \in \mathbb{R} \to \alpha_{tv}^*(\omega)$. That (4.37) holds follows from the identity (4.21). For the sequence $\{z_n\}_{n\in\mathbb{N}} := \{i\,n\}_{n\in\mathbb{N}}$ we get for n sufficiently large

$$\alpha_{tv}^*(z_n) = \frac{n}{c_0}\left[\frac{1}{\sqrt{1 + \tau_0\,n}} - 1\right] \leq -\frac{1}{2\,c_0}n.$$

Thus (4.40) holds.

The second part of Corollary 4.1 implies that K is not causal.

Model of Nachman, Smith and Waag

Theorem 4.7. *Let α_{nsw}^* as in (4.22). If*

$$\tilde{\tau}_m < \tau_m \qquad \text{for all} \qquad m \in \{1, \ldots, N\}, \tag{4.56}$$

then the kernel function K is causal.

Proof. Since for all $z \in \mathbb{C}_0$ and all $1 \leq m \leq N$, $1 - i\,\tau_m\,z$ does not vanish,

$$z \in \mathbb{C}_0 \to \alpha_{nsw}^*(z) = \frac{-i\,z}{c_0}\left[\frac{c_0}{\tilde{c}_0}\sqrt{\frac{1}{N}\sum_{m=1}^{N}\frac{1 - i\,\tilde{\tau}_m\,z}{1 - i\,\tau_m\,z}} - 1\right]$$

is the holomorphic extension of $\omega \in \mathbb{R} \to \alpha_{nsw}^*(\omega)$.

We use a similar notation as in Proposition 4.1.

$$z = \xi + i\,\eta = |z|\exp(i\,\phi) \in \mathbb{C}_0,$$

with some $\phi \in [0, \pi]$.

$$\alpha_{nsw}^*(z) = \frac{|z|}{c_0}\exp(i\,(\phi - \pi/2))\,(\Psi(z) - 1),$$

where

$$\Psi(z) = \sqrt{\sum_{m=1}^{N}\hat{\Psi}_m(z)} \quad \text{with} \quad \hat{\Psi}_m(z) := \frac{1}{N}\frac{c_0^2}{\tilde{c}_0^2}\frac{1 - i\,\tilde{\tau}_m\,z}{1 - i\,\tau_m\,z},$$

In the following we show that for all $z \in \mathbb{C}_0$

$$c_0 \alpha_{nsw}(z) = \eta(\mathrm{Re}(\Psi)(z) - 1) + \xi \mathrm{Im}(\Psi)(z) > 0, \tag{4.57}$$

which means that (4.38) holds. Then, according to Corollary 4.1 the function $t \to K(\mathbf{x}, t)$ is causal.

As in the proof of Proposition 4.1 we prove $\eta(\mathrm{Re}(\Psi)(z) - 1) > 0$ and $\xi \mathrm{Im}(\Psi)(z) > 0$.

- Taking into account (4.24) we define

$$s := \frac{1}{N} \frac{\tilde{c}_0^2}{\bar{c}_0^2} = \left(\sum_{m=1}^{N} \frac{\tilde{\tau}_m}{\tau_m} \right)^{-1}. \tag{4.58}$$

Using this notation, we get

$$
\begin{aligned}
\hat{\Psi}_m(z) &= s \frac{(1 + \tilde{\tau}_m \tau_m |z|^2) + \mathrm{i}\,(\tau_m \bar{z} - \tilde{\tau}_m z)}{|1 - \mathrm{i}\,\tau_m z|^2} \\
&= s \frac{\tilde{\tau}_m}{\tau_m} \frac{(\tau_m/\tilde{\tau}_m + \tau_m^2 |z|^2 + (\tau_m^2/\tilde{\tau}_m + \tau_m)\eta) + \mathrm{i}\,(\tau_m^2/\tilde{\tau}_m - \tau_m)\xi}{1 + \tau_m^2 |z|^2 + 2\tau_m \eta},
\end{aligned}
$$

Because $\tau_m/\tilde{\tau}_m > 1$, by assumption (4.56), it follows that for all $z \in \mathbb{C}_0$

$$\mathrm{Re}(\hat{\Psi}_m(z)) > s \frac{\tilde{\tau}_m}{\tau_m}.$$

Consequently, by using the definition of s, (4.58), it follows that

$$\mathrm{Re}\left(\sum_{m=1}^{N} \hat{\Psi}_m(z) \right) > s \sum_{m=1}^{N} \frac{\tilde{\tau}_m}{\tau_m} = 1.$$

Now, using (4.54) it follows

$$\mathrm{Re}(\Psi(z)) = \mathrm{Re}\left(\sqrt{\sum_{m=1}^{N} \Psi_m(z)} \right) \geq \sqrt{\mathrm{Re}\left(\sum_{m=1}^{N} \Psi_m(z) \right)} > 1$$

and consequently $\eta(\mathrm{Re}(\Psi(z)) - 1) > 0$.
- We have

$$\mathrm{Im}(\hat{\Psi}_m(z)) = s \frac{\tilde{\tau}_m}{\tau_m} \frac{(\tau_m^2/\tilde{\tau}_m - \tau_m)\xi}{1 + \tau_m^2 |z|^2 + 2\tau_m \eta}$$

together with the assumption (4.56), which state that $\tau_m/\tilde{\tau}_m > 1$, it follows that $\mathrm{sgn}(\mathrm{Im}(\hat{\Psi}_m(z))) = \mathrm{sgn}(\xi)$. According to our assumption, we take that complex root, such that the real part of the argument is non-negative which together with property (4.102) in the Appendix implies

$$\mathrm{sgn}\left(\mathrm{Im}\left(\sqrt{\sum_{m=1}^{N}\hat{\Psi}_m(z)}\right)\right)=\mathrm{sgn}\left(\mathrm{Im}\left(\sum_{m=1}^{N}\hat{\Psi}_m(z)\right)\right).$$

Therefore,

$$\mathrm{sgn}(\mathrm{Im}(\Psi(z)))=\mathrm{sgn}\left(\mathrm{Im}\left(\sqrt{\sum_{m=1}^{N}\hat{\Psi}_m(z)\}}\right)\right)$$

$$=\mathrm{sgn}\left(\mathrm{Im}\left(\sum_{m=1}^{N}\hat{\Psi}_m(z)\right)\right)=\mathrm{sgn}(\xi).$$

This shows the assertion.

Our Model

Theorem 4.8. *For α_0, $\tau_0 > 0$ and $\gamma \in (1,2]$ let α_{ksb}^* be defined as in (4.28). Then K, as defined in (4.39), is causal.*

Proof. The function $\hat{z} \to 1 + (-i\,\tau_0\hat{z})$ does not vanish in $\mathbb{C}_0$. Thus the holomorphic extension of $\omega \to \alpha_{ksb}^*(\omega)$ is given by

$$\hat{z} \in \mathbb{C}_0 \to \alpha_{ksb}^*(\hat{z}) = \frac{\alpha_0(-i\,\hat{z})}{c_0\sqrt{1 + (-i\,\tau_0\hat{z})^{\gamma-1}}}.$$

In the following let $\hat{z} \in \mathbb{C}_0$. For proving (4.38) we make a variable transformation

$$\alpha_{ksb}^*(\hat{z}) = \frac{-i\,\tau_0\hat{z}}{\sqrt{1 + (-i\,\tau_0\hat{z})^{\gamma-1}}} = \frac{\alpha_0}{\tau_0 c_0}\frac{-i\,z}{\sqrt{1 + (-i\,z^{\gamma-1})}},$$

and define

$$\Psi(z) = \frac{1}{\sqrt{\hat{\Psi}(z)}} \quad \text{and} \quad \hat{\Psi}(z) = 1 + (-i\,z)^{\gamma-1}.$$

Then, with this notation, in order to prove causality of K, it suffices to prove that for all $z \in \mathbb{C}_0$

$$\frac{\tau_0 c_0}{\alpha_0}\alpha_{ksb}(\hat{z}) = \eta\mathrm{Re}(\Psi(z)) + \xi\mathrm{Im}(\Psi(z)) \geq 0. \tag{4.59}$$

As in the proof of Proposition 4.1 we show that both terms $\eta\mathrm{Re}(\Psi(z))$ and $\xi\mathrm{Im}(\Psi(z))$ are non-negative, and then from Corollary 4.1 the assertion follows.

In order to prove (4.59) we note that the function $\hat{\Psi}$ here is the same as in (4.50) in the proof of Proposition 4.1 when $\tilde{\alpha}_0$ is set to $1/2$. Thus we can already rely on the series of manipulations for $\hat{\Psi}$ developed in the proof of Proposition 4.1.

- Since $\eta \geq 0$ it suffices to show that $\text{Re}(\Psi(z)) \geq 0$. We note that for a complex number $a + i\,b$

$$\text{Re}\left(\frac{1}{a+i\,b}\right) = \text{Re}\left(\frac{a-i\,b}{a^2+b^2}\right) = \frac{1}{a^2+b^2}\text{Re}(a+i\,b)\,.$$

Taking into account the definition of Ψ it therefore suffices to show that $\text{Re}\left(\sqrt{\hat{\Psi}(z)}\right) \geq 0$ in $\mathbb{C}_0$. Since $\text{Re}\left(\hat{\Psi}(z)\right) \geq 0$ in $\mathbb{C}_0$ for $\gamma \in (1,2]$, it follows that $\text{Re}\left(\sqrt{\hat{\Psi}(z)}\right) \geq 0$ in $\mathbb{C}_0$.

- Now, using that

$$\text{Im}\left(\frac{1}{a+i\,b}\right) = \text{Im}\left(\frac{a-i\,b}{a^2+b^2}\right) = -\frac{1}{a^2+b^2}\text{Im}(a+i\,b)\,,$$

it suffices to show that $-\xi\text{Im}(\sqrt{\hat{\Psi}(z)}) \geq 0$ for proving that $\xi\text{Im}(\Psi(z)) \geq 0$. The proof is along the lines as the analogous part in Proposition 4.1 by taking into account that here $\gamma \in (1,2)$ (in Proposition 4.1 $\gamma \in (0,1)$). In this case we have now that sign of δ is exactly opposite as in the proof of Proposition 4.1, which in turn gives that $\text{Im}(\Psi(z))$ has the opposite sign as well, and consequently $-\xi\text{Im}(\sqrt{\hat{\Psi}(z)}) \geq 0$. Thus the assertion follows from Corollary 4.1.

In experiments it has been discovered that several biological tissues satisfy a frequency power law (4.16) with exponent $\gamma \in (1,2)$ (cf. [5,45]). However, as it has been shown in Theorem 4.3, such models are not causal. Our proposed model approximates the frequency power law for small frequencies, which is actually the range where it has been experimentally validated. So, our proposed model, is valid in the actual range of experimentally measured data and extrapolates the measured data in a causal way. Figure 4.4 shows a comparison of α_{pl} and α_{ksb} in an experimental frequency range.

Model of Greenleaf and Patch

Proposition 4.2. *For $\alpha_0 > 0$ let α_{gp}^* be defined as in (4.25) with the specified values $\gamma \in \{1,2\}$. Then K, as defined in (4.39), is not causal.*

Proof. For the two specified models we have $\alpha_{gp1}(\omega) = a_0|\omega| > 0$ and $\alpha_{gp2}(\omega) = a_0\omega^2$. The respective holomorphic extensions are given by (cf. Proof of Theorem 4.5 and Theorem 4.3)

$$z \in \mathbb{C}_0 \mapsto \hat{\alpha}_{gp1}^*(z) := a_0 z + i\,\frac{2a_0}{\pi}z\log\left(\frac{z}{\omega_0}\right) \qquad (\omega_0 \neq 0,\,\text{fixed})$$

and

$$z \in \mathbb{C}_0 \mapsto \hat{\alpha}_{gp2}^*(z) := a_0\,(-i\,z)^2\,.$$

The assertion for $\gamma = 1$ follows as in the proof of Theorem 4.5 and the second assertion follows from Theorem 4.3 for $\gamma = 2$.

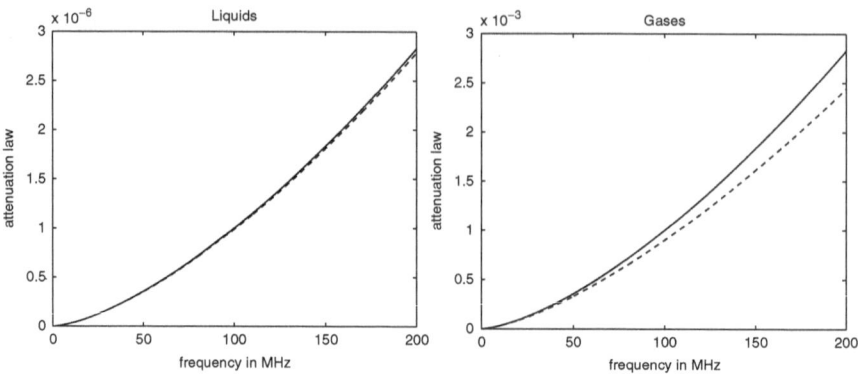

Fig. 4.4. For $\gamma = 1.5$: Comparison $\omega \to \alpha_{ksb}(\omega)$ (as defined in (4.28)) where $\alpha_0 := 2c_0\tau_0/\left|\cos(\frac{\pi}{2}\gamma)\right|$ (*dashed line*) and the power law $\alpha_{pl}(\omega) = |\tau_0\,\omega|^{\gamma}$ (as defined in (4.16)). For liquids: $\tau_0 = 10^{-6}$ MHz (*left picture*) and for gases: $\tau_0 = 10^{-4}$ MHz (*right picture*) (cf. [21]). Experiments for determining the power law coefficient are performed in the range $0 - 60$ MHz (cf. e.g. [40]), which is the basis for the range of the represented data

Model of Chen and Holm

Theorem 4.9. *Let $0 < \alpha_1 < 1$, $\gamma \in (0,2)$ and $\mathcal{G}$ as in (4.26). Then there does not exist a constant $c > 0$ such that*

$$supp(\mathcal{G}\,(\cdot,t)) \subseteq B_{ct}(\mathbf{0}) \qquad for \qquad t > 0, \tag{4.60}$$

i.e. for each $c > 0$ the function $t \mapsto \mathcal{G}(\mathbf{x}, t + |\mathbf{x}|/c)$ is not causal.

Proof. Let $t > 0$ be fixed. Assume that $\mathbf{x} \mapsto \mathcal{G}(\mathbf{x}, t)$ has support in $B_{ct}(\mathbf{0})$ for some $c > 0$. Then according to the Paley-Wiener-Schwartz Theorem (Cf. [12, 17]) the map $\mathbf{k} \mapsto \mathcal{F}^{-1}\{\mathcal{G}\}(\mathbf{k}, t)$ is infinitely differentiable. We show that this is not possible. According to (4.26) and (4.27), we have

$$\mathcal{F}_{3D}^{-1}\{\mathcal{G}\}(\mathbf{k}, t) = \frac{H(t)\,c_0^2}{(2\,\pi)^{3/2}} \exp\left(A(\mathbf{k})\,t\right) \frac{\sin(B(\mathbf{k})\,t)}{B(\mathbf{k})}$$

with

$$A(\mathbf{k}) := -\alpha_1\,c_0\,|\mathbf{k}|^{\gamma}\,, \qquad B(\mathbf{k}) := c_0\,\sqrt{|\mathbf{k}|^2 - \alpha_1^2\,|\mathbf{k}|^{2\,\gamma}}\,.$$

Since $\gamma \in (0,2)$, the function $\mathbf{k} \mapsto \exp\left(A(\mathbf{k})\,t\right)$ is not infinitely often differentiable at $\mathbf{k} = \mathbf{0}$ and since the holomorphic function $\frac{\sin(B(\mathbf{k})\,t)}{B(\mathbf{k})}$ does not vanish at $\mathbf{k} = \mathbf{0}$, it follows that $\mathbf{k} \mapsto \mathcal{F}_{3D}^{-1}\{\mathcal{G}\}(\mathbf{k}, t)$ is not infinitely often differentiable at $\mathbf{k} = \mathbf{0}$. Consequently, $\mathbf{x} \mapsto \mathcal{G}(\mathbf{x}, t)$ cannot have compact support, which concludes the proof.

4.5 Integro-Differential Equations Describing Attenuation

In the following we derive the integro-differential equations for the attenuated pressure p_{att} for various attenuation laws. Thereby, we first derive equations which the according attenuated Green functions $\mathcal{G}$ (cf. (4.9)) are satisfying, and then, by convolution, we derive the equations for p_{att}. The integro-differential equations are general in the sense, that they apply to arbitrary source terms f, and in particular to the source term f (4.6) of the forward problem of photoacoustic imaging with attenuated waves.

For this purpose, we rewrite $\nabla^2 \mathcal{F}\{\mathcal{G}\}$ by using its definition (4.9), i.e. $\mathcal{G} = K *_t \mathcal{G}_0$, and the product differentiation rule, which gives

$$\frac{1}{\sqrt{2\pi}} \nabla^2 \mathcal{F}\{\mathcal{G}\}$$
$$= \nabla^2 \mathcal{F}\{K\} \cdot \mathcal{F}\{\mathcal{G}_0\} + 2\nabla\mathcal{F}\{K\} \cdot \nabla\mathcal{F}\{\mathcal{G}_0\} + \mathcal{F}\{K\} \cdot \nabla^2\mathcal{F}\{\mathcal{G}_0\}. \quad (4.61)$$

To evaluate the expression on the right hand side, we calculate $\nabla\mathcal{F}\{K\}$ and $\nabla^2\mathcal{F}\{K\}$. From (4.10), it follows that

$$\nabla\mathcal{F}\{K\} = -\beta^{*\prime} \cdot \mathcal{F}\{K\} \cdot \text{sgn}, \quad (4.62)$$

where $\beta^{*\prime}$ denotes the derivative of $\beta^*(r, \omega)$ (cf. (4.10)) with respect to r. This together with the formula (4.103) in the Appendix implies that

$$\nabla^2\mathcal{F}\{K\}$$
$$= -\nabla \cdot (\beta^{*\prime} \cdot \mathcal{F}\{K\} \cdot \text{sgn})$$
$$= -(\nabla \cdot \text{sgn}) \cdot \beta^{*\prime} \cdot \mathcal{F}\{K\} - (\text{sgn} \cdot \nabla\beta^{*\prime}) \cdot \mathcal{F}\{K\} - (\text{sgn} \cdot \nabla\mathcal{F}\{K\}) \cdot \beta^{*\prime}$$
$$= \left[-\frac{2}{|\mathbf{x}|} \cdot \beta^{*\prime} - \beta^{*\prime\prime} + (\beta^{*\prime})^2 \right] \cdot \mathcal{F}\{K\}. \quad (4.63)$$

Inserting (4.62) and (4.63) into (4.61) and using again the identity $\mathcal{G} = K *_t \mathcal{G}_0$, shows that

$$\frac{1}{\sqrt{2\pi}} \nabla^2 \mathcal{F}\{\mathcal{G}\}$$
$$= \frac{1}{\sqrt{2\pi}} \left[-\frac{2}{|\mathbf{x}|} \cdot \beta^{*\prime} - \beta^{*\prime\prime} + (\beta^{*\prime})^2 \right] \cdot \mathcal{F}\{\mathcal{G}\} \quad (4.64)$$
$$- 2\beta^{*\prime} \cdot \mathcal{F}\{K\} \cdot (\text{sgn} \cdot \nabla\mathcal{F}\{\mathcal{G}_0\}) + \mathcal{F}\{K\} \cdot \nabla^2\mathcal{F}\{\mathcal{G}_0\}.$$

From this identity, together with the two following properties of $\mathcal{G}_0$,

$$\nabla\mathcal{F}\{\mathcal{G}_0\} = \left[\frac{i\omega}{c_0} - \frac{1}{|\mathbf{x}|} \right] \cdot \mathcal{F}\{\mathcal{G}_0\} \cdot \text{sgn}, \quad (4.65)$$

and

$$\nabla^2 \mathcal{F}\{\mathcal{G}_0\} + \frac{\omega^2}{c_0^2} \mathcal{F}\{\mathcal{G}_0\} = -\frac{1}{\sqrt{2\pi}} \delta_{\mathbf{x}}, \tag{4.66}$$

it follows that

$$\nabla^2 \mathcal{F}\{\mathcal{G}\} = \left[-\frac{2}{|\mathbf{x}|} \cdot \beta^{*\prime} - \beta^{*\prime\prime} + (\beta^{*\prime})^2 \right] \cdot \mathcal{F}\{\mathcal{G}\} - 2\left[\frac{i\omega}{c_0} - \frac{1}{|\mathbf{x}|} \right] \cdot \beta^{*\prime} \cdot \mathcal{F}\{\mathcal{G}\}$$
$$- \frac{\omega^2}{c_0^2} \cdot \mathcal{F}\{\mathcal{G}\} - \mathcal{F}\{K\} \cdot \delta_{\mathbf{x}}. \tag{4.67}$$

Inserting the identity $\mathcal{F}\{K\}(\mathbf{x}, \omega) \cdot \delta_{\mathbf{x}} = \mathcal{F}\{K\}(\mathbf{0}, \omega) \cdot \delta_{\mathbf{x}}$ in (4.67) gives the *Helmholtz equation*

$$\nabla^2 \mathcal{F}\{\mathcal{G}\} - \left[\beta^{*\prime} + \frac{(-i\,\omega)}{c_0} \right]^2 \cdot \mathcal{F}\{\mathcal{G}\} = -\beta^{*\prime\prime} \cdot \mathcal{F}\{\mathcal{G}\} - \mathcal{F}\{K\}(\mathbf{0}, \cdot) \cdot \delta_{\mathbf{x}}$$
$$= -\beta^{*\prime\prime} \cdot \mathcal{F}\{\mathcal{G}\}$$
$$- \frac{1}{\sqrt{2\pi}} \exp\left(-\beta^*(\mathbf{0}, \omega)\right) \cdot \delta_{\mathbf{x}}. \tag{4.68}$$

To reformulate (4.68) in space–time coordinates, we introduce two convolution operators:

$$D_* f := K_* *_t f \qquad \text{and} \qquad D'_* f := K'_* *_t f, \tag{4.69}$$

where the kernels K_* and K'_* are given by

$$K_* := K_*(\mathbf{x}, t) := K_*(|\mathbf{x}|, t) \quad \text{and} \quad K_*(r, t) := \frac{1}{\sqrt{2\pi}} \mathcal{F}^{-1}\{\beta^{*\prime}\}(r, t) \tag{4.70}$$

and

$$K'_* := K'_*(\mathbf{x}, t) := K'_*(|\mathbf{x}|, t) \quad \text{and} \quad K'_*(r, t) = \frac{1}{\sqrt{2\pi}} \mathcal{F}^{-1}\{\beta^{*\prime\prime}\}(r, t). \tag{4.71}$$

Using these operators and applying the inverse Fourier transform to (4.68) gives

$$\nabla^2 \mathcal{G} - \left[D_* + \frac{1}{c_0} \frac{\partial}{\partial t} \right]^2 \mathcal{G} = -D'_* \mathcal{G} - K(\mathbf{0}, \cdot)\delta_{\mathbf{x}}. \tag{4.72}$$

In the case that $\beta^*(|\mathbf{x}|, \omega) = \alpha^*(\omega)|\mathbf{x}|$ is of standard form (4.13), it follows that

$$K_*(t) = \frac{1}{\sqrt{2\pi}} \mathcal{F}^{-1}\{\alpha^*\}(t) \qquad \text{and} \qquad K'_* \equiv 0. \tag{4.73}$$

For a general source term f, we denote the attenuated wave by p_{att}. That is

$$p_{\text{att}} := p_{\text{att}}(\mathbf{x}, t) = \mathcal{G} *_{\mathbf{x}, t} f =: \mathcal{A}f,$$

where $\mathcal{A}$ is the convolution operator according to the Green function $\mathcal{G}$. This then shows that p_{att} satisfies the integro-differential equation

$$\nabla^2 p_{\text{att}} - \frac{1}{c_0^2} \frac{\partial^2 p_{\text{att}}}{\partial t^2} = -\mathcal{A}_s f \, , \tag{4.74}$$

where $\mathcal{A}_s$ denotes the space–time convolution operator with kernel

$$K_s := K_s(\mathbf{x}, t) := -(\mathcal{B}\mathcal{G})(\mathbf{x}, t) + (D'_* \mathcal{G})(\mathbf{x}, t) + K(\mathbf{0}, t) \cdot \delta_{\mathbf{x}}(\mathbf{x}) \tag{4.75}$$

and

$$\mathcal{B} := D_*^2 + \frac{2}{c_0} D_* \frac{\partial}{\partial t} \, . \tag{4.76}$$

Equation (4.74) is called *pressure wave equation with attenuation coefficient* β^*. We emphasize that β^* determines the operators D_* and D'_* which in turn determine the operator $\mathcal{A}_s$, which in turn determines p_{att} - this reveals the dependence of p_{att} from β_*.

Remark 4.4. Let $\beta^*(r, \omega) = \alpha^*(\omega)r$ be the standard attenuation model (cf. (4.12)). Assuming that the associated kernel K (cf. (4.39)) is causal, it follows that

$$|\nabla K| = \frac{1}{\sqrt{2\pi}} \left| \mathcal{F}^{-1} \{ \alpha^* \cdot \exp(-\alpha^* |\mathbf{x}|) \} \right| \, .$$

Using some sequence $\{\mathbf{x}_n\}$ satisfying $\mathbf{x}_n \neq \mathbf{0}$ and $\mathbf{x}_n \to \mathbf{0}$ shows that

$$\lim_{n \to \infty} |\nabla K| (\mathbf{x}_n, t) = \frac{1}{\sqrt{2\pi}} \left| \mathcal{F}^{-1} \{ \alpha^* \} (t) \right| \underbrace{=}_{(4.73)} |K_*(t)| \, .$$

Due to the causality of K the left hand side is zero for $t < 0$, and thus K_* is also causal.

Because the convolution of causal distributions is well-defined, the operator D_* is well-defined on all causal distributions. Moreover, since $K'_* = 0$, it follows that $D'_* \equiv 0$. Using that K_* depends only on t it follows that

$$(D_* \mathcal{G}) *_{\mathbf{x},t} f = [K_* *_t \mathcal{G}] *_{\mathbf{x},t} f = K_* *_t [\mathcal{G} *_{\mathbf{x},t} f] = D_*(\mathcal{G} *_{\mathbf{x},t} f).$$

Convolving each term in (4.72) with a function f, using the previous identity and that $D'_* \equiv 0$, it follows that

$$\nabla^2 p_{\text{att}} - \left[D_* + \frac{1}{c_0} \frac{\partial}{\partial t} \right]^2 p_{\text{att}} = -f \tag{4.77}$$

where

$$D_* \cdot = \frac{1}{\sqrt{2\pi}} \mathcal{F} \{ \alpha^* \} (t) *_t \cdot \, . \tag{4.78}$$

In the following we derive the common forms of the wave equation models corresponding to the various attenuation models listed in Sect. 4.2.

Power Laws

- Let $0 < \gamma \notin \mathbb{N}$ and $0 < \alpha_0$. We note that the *Riemann-Liouville fractional derivative* with respect to time, denote by D_t^γ (see [20, 34]), is defined in the Fourier domain by

$$\mathcal{F}\{D_t^\gamma f\} = (-\mathrm{i}\omega)^\gamma \mathcal{F}\{f\}, \tag{4.79}$$

and satisfies

$$D_t^{2\gamma} f = D_t^\gamma D_t^\gamma f \quad \text{and} \quad \frac{\partial}{\partial t} D_t^\gamma f = D_t^\gamma \frac{\partial}{\partial t} f = D_t^{\gamma+1} f. \tag{4.80}$$

From this together with (4.15) and (4.78), we infer

$$D_* = \tilde{\alpha}_0\, D_t^\gamma \quad \text{with} \quad \tilde{\alpha}_0 := \frac{\alpha_0}{\cos(\pi\,\gamma/2)} \tag{4.81}$$

and thus wave equation (4.77) reads as follows

$$\nabla^2 p_{\text{att}} - \left[\tilde{\alpha}_0\, D_t^\gamma + \frac{1}{c_0}\frac{\partial}{\partial t}\right]^2 p_{\text{att}} = -f. \tag{4.82}$$

- Let $\gamma = 1$, then for the frequency power law (4.17) it follows from the Fourier transform table I in [28]

$$D_* = -\frac{4\,\alpha_0}{2\,\pi} \left[\frac{H(t)}{t^2} - (\log|\omega_0|)\,\delta_t'\right] *_t$$

$$= -\frac{4\,\alpha_0}{2\,\pi}\frac{H(t)}{t^2} *_t \; + \frac{4\,\alpha_0}{\sqrt{2\,\pi}}(\log|\omega_0|)\frac{\partial}{\partial t}.$$

Szabo's Attenuation Law:

Let $0 < \alpha_0$ and $0 < \gamma \notin \mathbb{N}$. From (4.19) and (4.79), we get

$$\left[D_* + \frac{1}{c_0}\frac{\partial}{\partial t}\right]^2 = \frac{1}{c_0^2}\frac{\partial^2}{\partial t^2} + \frac{2\,\tilde{\alpha}_0}{c_0}\frac{\partial}{\partial t} D_t^\gamma$$

and thus wave equation (4.77) reads as follows

$$\nabla^2 p_{\text{att}} - \frac{1}{c_0^2}\frac{\partial^2 p_{\text{att}}}{\partial t^2} - \frac{2\,\tilde{\alpha}_0}{c_0}\frac{\partial}{\partial t} D_t^\gamma p_{\text{att}} = -f(\mathbf{x}, t). \tag{4.83}$$

Thermo-Viscous Attenuation Law:

From (4.20) we get

$$\left(\mathrm{Id} + \tau_0\frac{\partial}{\partial t}\right)\left[D_* + \frac{1}{c_0}\frac{\partial}{\partial t}\right]^2 = \frac{1}{c_0^2}\frac{\partial^2}{\partial t^2}$$

and thus (4.77) becomes

$$\left(\mathrm{Id} + \tau_0 \frac{\partial}{\partial t}\right) \nabla^2 p_{\mathrm{att}} - \frac{1}{c_0^2} \frac{\partial^2 p_{\mathrm{att}}}{\partial t^2} = -\left(\mathrm{Id} + \tau_0 \frac{\partial}{\partial t}\right) f. \tag{4.84}$$

This equation is called the *thermo-viscous wave equation*.

Nachman, Smith and Waag [29]:

We carry out the details only for one relaxation process.

- $N = 1$: From (4.22) we get

$$\left(\alpha^*(\omega) + \frac{(-\mathrm{i}\,\omega)}{c_0}\right)^2 = \frac{(-\mathrm{i}\,\omega)^2}{\tilde{c}_0^2} \frac{1 - \mathrm{i}\,\tilde{\tau}_1\,\omega}{1 - \mathrm{i}\,\tau_1\,\omega}$$

which implies

$$\left(\mathrm{Id} + \tau_1 \frac{\partial}{\partial t}\right) \left[D_* + \frac{1}{c_0}\frac{\partial}{\partial t}\right]^2 = \left(\mathrm{Id} + \tilde{\tau}_1 \frac{\partial}{\partial t}\right) \frac{1}{\tilde{c}_0^2} \frac{\partial^2}{\partial t^2}.$$

Thus (4.77) reads as follows

$$\left(\mathrm{Id} + \tau_1 \frac{\partial}{\partial t}\right) \nabla^2 p_{\mathrm{att}} - \frac{1}{\tilde{c}_0^2}\left(\mathrm{Id} + \tilde{\tau}_1 \frac{\partial}{\partial t}\right) \frac{\partial^2 p_{\mathrm{att}}}{\partial t^2} = -\left(\mathrm{Id} + \tau_1 \frac{\partial}{\partial t}\right) f.$$
$$\tag{4.85}$$

If the term with $\tilde{\tau}_1 = 0$ is dropped and $\tilde{c}_0$ is replaced by c_0, then we obtain the thermo-viscous wave equation (4.84).
- $N > 1$: For the general case we refer to equation (26) in [29].

Greenleaf and Patch [32]:

- For $\gamma = 2$ the attenuation coefficient equals to

$$\alpha^*(\omega) = \alpha_0\,\omega^2 = \tilde{\alpha}_0\,(-\mathrm{i}\,\omega)^2,$$

where $\tilde{\alpha}_0$ is defined as in (4.81) and thus

$$D_* = -\alpha_0 \frac{\partial^2}{\partial t^2} = -\alpha_0\, D_t^2,$$

which gives wave equation (4.82) with $\gamma = 2$.
- For $\gamma = 1$ we have

$$\alpha^*(\omega) = \alpha_0\,|\omega| = \alpha_0\,(-\mathrm{i}\,\omega)\,\mathrm{i}\,\mathrm{sgn}(\omega)$$

and thus

$$D_* = \alpha_0\,\mathbf{D}^{-1} = -\alpha_0 \frac{\partial}{\partial t}\mathcal{H},$$

where $\mathbf{D}^{-1}$ and $\mathcal{H}$ denote the *Riesz fractional differentiation operator* and the Hilbert transform (cf. Appendix), respectively. Therefore the wave equation reads as follows

$$\nabla^2 p_{\text{att}} - \left[\alpha_0 \, \mathbf{D}^{-1} + \frac{1}{c_0} \frac{\partial}{\partial t}\right]^2 p_{\text{att}} = -f \, .$$

Chen and Holm [6]:

Let $\gamma \in (0,2)$. The Green function defined by (4.26) satisfies the Helmholtz equation

$$\frac{\partial^2 \mathcal{F} \{\mathcal{G}\}}{\partial t^2}(\mathbf{k}, t) + 2\,\alpha_1 \, c_0 \, |\mathbf{k}|^\gamma \, \frac{\partial \mathcal{F} \{\mathcal{G}\}}{\partial t}(\mathbf{k}, t) + c_0^2 \, |\mathbf{k}|^2 \, \mathcal{F} \{\mathcal{G}\}(\mathbf{k}, t)$$
$$= \frac{c_0^2}{(2\,\pi)^{3/2}} \, \delta(t) \tag{4.86}$$

for $t \in \mathbb{R}$ and $\mathbf{k} \in \mathbb{R}^3$. Since the fractional Laplacian for a rotational symmetric function f and $\gamma \in (0,2)$ is defined by (cf. Definition (2.10.1) in ([20])

$$(-\nabla^2)^{\gamma/2} f(\mathbf{x})$$
$$:= \frac{1}{\sqrt{(2\,\pi)^3}} \int_{\mathbb{R}^3} \exp\left(\mathbf{x} \cdot \mathbf{k}\right) \left[|\mathbf{k}|^\gamma \, \frac{1}{\sqrt{(2\,\pi)^3}} \int_{\mathbb{R}^3} \exp\left(-\mathbf{x} \cdot \mathbf{k}\right) f(\mathbf{x}) \, \mathrm{d}\mathbf{x}\right] \mathrm{d}\mathbf{k}$$
$$= \mathcal{F}_{3D} \left\{|\mathbf{k}|^\gamma \, \mathcal{F}_{3D}^{-1}\{f\}(\mathbf{k})\right\}(\mathbf{x}) \, ,$$

we obtain the following wave equation for $p_{\text{att}} := \mathcal{G} *_{\mathbf{x}, t} f$

$$\nabla^2 p_{\text{att}} - \frac{1}{c_0^2} \frac{\partial^2 p_{\text{att}}}{\partial t^2} - \frac{2\,\alpha_1}{c_0} \frac{\partial}{\partial t} \, (-\nabla^2)^{\gamma/2} \, p_{\text{att}} = -f(\mathbf{x}, t) \, . \tag{4.87}$$

We note that Chen and Holm used instead of $2\,\alpha_1/c_0$ the term $2\,\alpha_1/c_0^{1-\gamma}$.

Our Model [22]:

From (4.28) we get

$$\left(\mathrm{Id} + \tau_0^{\gamma-1} \, D_t^{\gamma-1}\right) \left[D_* + \frac{1}{c_0} \frac{\partial}{\partial t}\right]^2 = \frac{1}{c_0^2} \frac{\partial^2}{\partial t^2} \left(\alpha_0 \, \mathrm{Id} + L^{1/2}\right)^2$$

where the time convolution operator $L^{1/2}$ is the convolution operator with kernel

$$l(t) := L^{1/2}(\delta_t) = \frac{1}{\sqrt{2\,\pi}} \mathcal{F} \left\{\sqrt{1 + (-\mathrm{i}\,\tau_0 \, \omega)^{\gamma-1}}\right\} \, .$$

Consequently, $L := (L^{1/2})^2 = \mathrm{Id} + \tau_0 \, D_t^{\gamma-1}$ and (4.77) can be rewritten as follows

$$\left(\mathrm{Id} + \tau_0^{\gamma-1} \, D_t^{\gamma-1}\right) \nabla^2 p_{\mathrm{att}} - \frac{1}{c_0^2} \left(\alpha_0 \, \mathrm{Id} + L^{1/2}\right)^2 \frac{\partial^2 p_{\mathrm{att}}}{\partial t^2}$$

$$= -\left(\mathrm{Id} + \tau_0^{\gamma-1} \, D_t^{\gamma-1}\right) f. \tag{4.88}$$

4.6 Pressure Relation

In this section we derive the relation between p_{att} and p_0 when the source term is of the form (4.6). This chapter is a special instance of Sect. 4.5. However, utilizing the special structure of the source term different formulas can be derived.

Attenuation is defined as a multiplicative law (in the frequency domain) relating the amplitudes of an attenuated and an unattenuated wave initialized by a delta impulse. Here we are concerned in deriving the convolution relation between the solution p_0 of (4.1) (or equivalently of (4.4) and (4.5)) and the attenuated wave function p_{att}, which, according to (4.9) and (4.29), is given by

$$p_{\mathrm{att}} = \mathcal{G} *_{\mathbf{x},t} f = (K *_t \mathcal{G}_0) *_{\mathbf{x},t} f, \tag{4.89}$$

with f from (4.6). Using (4.7) and the rotational symmetry of K, it follows that

$p_{\mathrm{att}}(\mathbf{x}, t)$

$$= \int_{\mathbb{R}} \int_{\mathbb{R}^3} (K *_t \mathcal{G}_0)(\mathbf{x} - \mathbf{x}', t - t'') \rho(\mathbf{x}') \, d\mathbf{x}' \frac{\partial \delta_t}{\partial t}(t'') \, dt''$$

$$= \int_{\mathbb{R}} \int_{\mathbb{R}^3} \int_{\mathbb{R}} K(\mathbf{x} - \mathbf{x}', t - t' - t'') \mathcal{G}_0(\mathbf{x} - \mathbf{x}', t') \, dt' \rho(\mathbf{x}') \, d\mathbf{x}' \frac{\partial \delta_t}{\partial t}(t'') \, dt''$$

$$= \int_{\mathbb{R}} \int_{\mathbb{R}^3} \int_{\mathbb{R}} \frac{\partial}{\partial t} K(\mathbf{x} - \mathbf{x}', t - t' - t'') \mathcal{G}_0(\mathbf{x} - \mathbf{x}', t') \rho(\mathbf{x}') \delta_t(t'') \, dt' \, d\mathbf{x}' \, dt''$$

$$= \int_{\mathbb{R}^3} \int_{\mathbb{R}} \frac{\partial}{\partial t} K(|\mathbf{x} - \mathbf{x}'|, t - t') \frac{\delta_t(t' - |\mathbf{x} - \mathbf{x}'| / c_0)}{4\pi |\mathbf{x} - \mathbf{x}'|} \rho(\mathbf{x}') \, dt' \, d\mathbf{x}'$$

$$= \int_{\mathbb{R}^3} \frac{\partial}{\partial t} K(|\mathbf{x} - \mathbf{x}'|, t - |\mathbf{x} - \mathbf{x}'| / c_0) \frac{\rho(\mathbf{x}')}{4\pi |\mathbf{x} - \mathbf{x}'|} \, d\mathbf{x}'.$$

Using the representation $\mathbf{x}' - \mathbf{x} = r' \mathbf{s}$ with $r' \geq 0$ and $\mathbf{s} \in S^2$, it follows that

$$p_{\mathrm{att}}(\mathbf{x}, t) = \frac{1}{4\pi} \int_0^\infty \frac{\partial}{\partial t} K(r', t - r'/c_0) \, r' \int_{S^2} \rho(\mathbf{x} + r'\mathbf{s}) \, d\mathbf{s} \, dr'. \tag{4.90}$$

Moreover,

$$
\begin{aligned}
p_0(\mathbf{x}, t) &= \frac{\partial}{\partial t} \int_{\mathbb{R}^3} \frac{\delta_t(t - |\mathbf{x} - \mathbf{x}'|/c_0)}{4\pi |\mathbf{x} - \mathbf{x}'|} \rho(\mathbf{x}') \, d\mathbf{x}' \\
&= \frac{\partial}{\partial t} \int_0^\infty r'^2 \int_{S^2} \frac{\delta_t(t - r'/c_0)}{4\pi r'} \rho(\mathbf{x} + r'\mathbf{s}) \, ds \, dr' \\
&= \frac{\partial}{\partial t} \int_0^\infty r' \frac{\delta_t(t - r'/c_0)}{4\pi} \int_{S^2} \rho(\mathbf{x} + r'\mathbf{s}) \, ds \, dr' \\
&= \frac{\partial}{\partial t} \int_0^\infty \frac{c_0^2 r''}{4\pi} \delta_t(t - r'') \int_{S^2} \rho(\mathbf{x} + c_0 r''\mathbf{s}) \, ds \, dr'' \\
&= \frac{\partial}{\partial t} \left(\frac{c_0^2 t}{4\pi} \int_{S^2} \rho(\mathbf{x} + (c_0 t)\mathbf{s}) \, ds \right).
\end{aligned}
$$

This gives

$$
r' \int_{S^2} \rho(\mathbf{x} + r'\mathbf{s}) \, ds = \frac{4\pi}{c_0} \int_0^{r'/c_0} p_0(\mathbf{x}, t') \, dt'. \tag{4.91}
$$

Now, denoting

$$
F(t, r') := \int_0^{r'} \frac{\partial}{\partial t} K(r'', t - r''/c_0) \, dr'', \qquad G(r') := \int_0^{r'/c_0} p_0(\mathbf{x}, r'') \, dr''
$$

and

$$
F(t, \infty) = \lim_{r' \to \infty} F(t, r')
$$

it follows from (4.90) and (4.91) and the fact that $p_0(\mathbf{x}, 0) = 0$ that

$$
\begin{aligned}
p_{\text{att}}(\mathbf{x}, t) &= \frac{1}{c_0} \int_0^\infty F'(t, r') G(r') \, dr' \\
&= -\frac{1}{c_0} \int_0^\infty F(t, r') \underbrace{G'(r')}_{= p_0(\mathbf{x}, r'/c_0)/c_0} dr' + \frac{1}{c_0} F(t, r') G(r')|_{r'=0}^\infty \\
&= \frac{1}{c_0^2} \left(F(t, \infty) \int_0^\infty p_0(\mathbf{x}, r'/c_0) \, dr' - \int_0^\infty F(t, r') p_0(\mathbf{x}, r'/c_0) \, dr' \right) \\
&= \frac{1}{c_0} \left(F(t, \infty) \int_0^\infty p_0(\mathbf{x}, t') \, dt' - \int_0^\infty F(t, c_0 t') p_0(\mathbf{x}, t') \, dt' \right) \\
&=: \int_0^\infty \mathcal{M}(t, t') p_0(\mathbf{x}, t') \, dt',
\end{aligned}
$$

$$\tag{4.92}$$

where

$$
\mathcal{M}(t, t') := \frac{1}{c_0} \left(F(t, \infty) - F(t, c_0 t') \right). \tag{4.93}
$$

In the following we derive an equivalent representation of $\mathcal{M}$ in terms of the attenuation coefficient, under the assumption that the attenuation coefficient

α^* is such that K is causal. From (4.10), (4.12) and Item 4 in the Appendix, it follows that

$$K(r', t - r'/c_0) = \frac{1}{\sqrt{2\pi}} \mathcal{F}^{-1} \left\{ \exp\left(-\alpha^*(\omega)\, r' + i\, \frac{\omega}{c_0}\, r' \right) \right\}(t).$$

which implies

$$
\begin{aligned}
F(t, c_0\, t') \\
= \int_0^{c_0\, t'} \frac{\partial}{\partial t} K(r', t - r'/c_0)\, dr' \\
= \frac{1}{\sqrt{2\pi}} \mathcal{F}^{-1} \left\{ \frac{-i\,\omega\, [\exp(-\alpha^*(\omega)\, c_0\, t' + i\,\omega\, t') - 1]}{-\alpha^*(\omega) + i\,\omega/c_0} \right\}(t) \\
= \frac{1}{\sqrt{2\pi}} \mathcal{F}^{-1} \left\{ \frac{-i\,\omega}{-\alpha^*(\omega) + i\,\omega/c_0} \right\}(t) *_t [K(c_0\, t', t - t') - \delta_t(t)].
\end{aligned}
\tag{4.94}
$$

Since K is causal, it satisfies $K(c_0\, t', t - t') = 0$ for $t < t'$, and therefore

$$F(t, \infty) = \lim_{t' \to \infty} F(t, c_0\, t') = \frac{1}{\sqrt{2\pi}} \mathcal{F}^{-1} \left\{ \frac{-i\,\omega}{-\alpha^*(\omega) + i\,\omega/c_0} \right\}(t).$$

Hence (4.94) can be written as follows:

$$F(t, c_0\, t') = F(t, \infty) *_t K(c_0\, t', t - t') - F(t, \infty).$$

and therefore (4.93) simplifies to

$$
\begin{aligned}
\mathcal{M}(t, t') = -\frac{1}{c_0} F(t, \infty) *_t K(c_0\, t', t - t') \\
= \frac{1}{\sqrt{2\pi}} \mathcal{F}^{-1} \left\{ \frac{i\,\omega\, \exp(-\alpha^*(\omega)\, c_0\, t' + i\,\omega\, t')}{-\alpha^*(\omega)\, c_0 + i\,\omega} \right\}(t).
\end{aligned}
\tag{4.95}
$$

Note that $\mathcal{M}(t, 0) = -\frac{1}{c_0} F(t, \infty)$. The following lemma shows that if K is causal

$$\mathcal{M}(t, t') = 0 \qquad \text{if} \qquad 0 < t < t' \tag{4.96}$$

and therefore the upper limit of integration in the last term (4.92) can be replaced by t. This means that the set of attenuated pressure values

$$\{p_{\text{att}}(\mathbf{x}, s) : 0 \leq s \leq t\}$$

depend only on the unattenuated pressure values

$$\{p_0(\mathbf{x}, s) : 0 \leq s \leq t\}.$$

Lemma 4.3. *Let K from (4.10) be causal with $\beta^*(r, \omega) := \alpha^*(\omega)\, r$. Moreover, assume that $\alpha^*(\omega) \neq i\,\omega/c_0$ for every $\omega \in \mathbb{R}$, and let $\mathcal{M}$ be as defined in (4.95). Then:*

- *The function $t \to \mathcal{M}(t, 0)$ is causal.*
- *For every $t' > 0$*

$$\mathcal{M}(t, t') = 0 \text{ for all } t < |t'|. \tag{4.97}$$

Proof. • In order to prove causality of $t \to \mathcal{M}(t, 0)$ we verify the three assumptions of Theorem 4.2 for the tempered distribution

$$\mathcal{F}\{\mathcal{M}\}(\omega, 0) = \frac{1}{\sqrt{2\pi}} \frac{\omega}{k(\omega) c_0} \quad \text{with} \quad k(\omega) := \mathrm{i}\, \alpha^*(\omega) + \frac{\omega}{c_0}.$$

Since K is causal, as has been shown in Remark 4.4, also the function

$$t \mapsto K_*(t) = \frac{1}{\sqrt{2\pi}} \mathcal{F}^{-1}\{\alpha^*\}(t)$$

is causal. Now, using Theorem 4.2, it follows that:
(1) α_* is holomorphic in $\overset{\circ}{\mathbb{C}}_0$,
(2) $\alpha^*(\xi + \mathrm{i}\, \eta) \to \alpha^*(\xi)$ for $\eta \to 0$ in $\mathcal{S}'$ and
(3) for each $\epsilon > 0$ there exists a polynomial P such that $|\alpha^*(z)| \leq P(|z|)$ for $z \in \mathbb{C}_\epsilon$.
Since $\alpha_*(\omega) \neq \mathrm{i}\, \omega/c_0$ for all $\omega \in \mathbb{R}$ together with in [2, Theorem 2.7] it follows that $z \to \alpha_*(z)$ is unique holomorphic extension to $\mathbb{C}_0$ and therefore $z \to \alpha_*(z)$ cannot be identical to $z \to \mathrm{i}\, z/c_0$, the holomorphic extension of $\mathrm{i}\, \omega/c_0$. $\alpha_*(z) \neq \mathrm{i}\, z/c_0$ for $z \in \mathbb{C}_0$ implies that k has no zeros and hence $z/k(z)$ is holomorphic on $\overset{\circ}{\mathbb{C}}_0$. This shows that Item 1 in Theorem 4.2 is satisfied for $t \to \mathcal{M}(t, 0)$.
Since $1/(k(\xi + \mathrm{i}\, \eta)\, k(\xi))$ is bounded and $k(\xi + \mathrm{i}\, \eta) \to k(\xi)$ for $\eta \to 0$ in $\mathcal{S}'$, it follows that for all $\psi \in \mathcal{S}$

$$\lim_{\eta \to 0} \int_{\mathbb{R}} \left[\frac{k(\xi) - k(\xi + \mathrm{i}\, \eta)}{k(\xi + \mathrm{i}\, \eta)\, k(\xi)} \right] \psi(\xi)\, \mathrm{d}\xi \to 0$$

i.e. $1/k(\xi + \mathrm{i}\, \eta) \to 1/k(\xi)$ for $\eta \to 0$ in $\mathcal{S}'$. Hence $\mathcal{F}\{\mathcal{M}\}(\xi + \mathrm{i}\, \eta, 0) \to \mathcal{F}\{\mathcal{M}\}(\xi, 0)$ for $\eta \to 0$ in $\mathcal{S}'$, which shows Item 2 in Theorem 4.2.
Since $k(z)$ does not vanish on $\mathbb{C}_0$ and $|k(z)|$ is bounded by a polynomial in $|z|$ for $z \in \overset{\circ}{\mathbb{C}}_0$, it follows that $|z/k(z)|$ is bounded by a polynomial in $|z|$. Hence Item 3 in Theorem 4.2 is satisfied and consequently $t \to \mathcal{M}(t, 0)$ is causal.
- Property (4.97) is satisfied if

$$\mathcal{M}(t + |t'|, t') = 0 \qquad \text{for} \quad t < 0. \tag{4.98}$$

From (4.95) and

$$K(c_0\, t', t) = \mathcal{F}^{-1} \left\{ \frac{e^{-\alpha^*(\omega)\, c_0\, t'}}{\sqrt{2\pi}} \right\}(t),$$

it follows that

$$\mathcal{M}(t + |t'|, t') = \mathcal{M}(t, 0) *_t K(c_0 t', t).$$

Since $t \mapsto \mathcal{M}(t, 0)$ and $t \mapsto K(c_0 t', t)$ are causal, their convolution is also causal (cf. Item 7 in the Appendix). This proves property (4.98) and concludes the proof.

Remark 4.5. Assume that the attenuation coefficient is given by

$$\alpha^*(\omega) = \mathrm{i}\,\omega/c_0 \qquad \text{for} \qquad \omega \in \mathbb{R}.$$

Then

$$K(r, t) = \frac{1}{\sqrt{2\,\pi}}\, \mathcal{F}^{-1}\left\{\exp\left(-\mathrm{i}\,\omega\, r/c_0\right)\right\}(t) = \delta(t + r/c_0)$$

which implies together with (4.9) and (4.7) that

$$\mathcal{G}(r, t) = \frac{\delta(t)}{4\,\pi\, r}.$$

But this function does not correspond to the intuition of an attenuated wave, which is manifested by the convolution equation (4.9), which should give a smooth decay of frequency components over travel distance. With this Green function $\mathcal{G}$ the input impulse collapses immediately and consequently, in this case, the assumption $\alpha^*(\omega) \neq \mathrm{i}\,\omega/c_0$ in Theorem 4.3 reflects physical reality.

4.7 Solution of the Integral Equation

The inverse problem of photoacoustics with attenuated waves reduces to solving the integral equation (4.92) for p_0, and to the standard photoacoustical inverse problem, which consists in calculating the initial pressure ρ in the wave equation (4.4) from measurements of $p_0(\mathbf{x}, t)$ over time on a manifold surrounding the object of interest. The standard photoacoustical imaging problem is not discussed here further, but we focus on the the integral equation (4.92).

In the following we investigate the ill–conditionness of the integral equation (4.92), where the kernel $\mathcal{M}$ is given from the attenuation law (4.28) with $\gamma \in (1, 2]$. In this case the model is causal and the parameter range $\gamma \in (1, 2]$ is relevant for biological imaging.

In order to estimate the ill–conditionness of the integral equation (4.92) it is rewritten in Fourier domain:

$$\mathcal{F}\left\{p_{\mathrm{att}}\right\}(\mathbf{x}_0, \omega) = \int_{\mathbb{R}} \mathcal{F}\left\{\mathcal{M}\right\}(\omega, t')\, p_0(\mathbf{x}_0, t')\, dt'. \tag{4.99}$$

After discretization the ill-conditionness of this equation is reflected by the decay rate of the singular values of the matrix $\mathcal{F}\left\{\mathcal{M}\right\}(\omega, t')$ at certain discrete frequencies and time instances.

We consider simple test examples of attenuation coefficients ρ (as in (4.3)), which are characteristic functions of balls with center at the origin and radiii R. For these examples we investigate the dependence of the ill–conditionedness of (4.99) on the radius R and the location $\mathbf{x}_0$. For applications in photoacoustic imaging $\mathbf{x}_0$ would be the location of a detector outside of the object of interest, to be imaged. Then, by solving the integral equation (4.99) p_0 can be calculated, and in turn, the absorption energy ρ can be reconstructed with standard backprojection formulas. Since

$$\text{supp}(p_0(\mathbf{x}_0, \cdot)) = [(|\mathbf{x}_0| - R)/c_0, (|\mathbf{x}_0| + R)/c_0] \,,$$

the integral (4.99) can be rewritten as

$$\mathcal{F}\{p_{att}\}(\mathbf{x}_0, \omega) = \int_{(R_0 - R)/c_0}^{\infty} \mathcal{F}\{\mathcal{M}\}(\omega, t') \, p_0(\mathbf{x}_0, t') \, dt' \,, \tag{4.100}$$

where $R_0 = |\mathbf{x}|_0$.

In the following we analyze the integral (4.100) in terms of the two parameters R and $R_0 = |\mathbf{x}_0|$. This gives a clue on the effect of attenuation in terms of the size of the object and the distance of the location $\mathbf{x}_0$ to the simple object. In order to show the effect of attenuation on the single frequencies, we make a singular value decomposition of the kernel of the integral (4.100).

Example 2. For small frequencies the attenuation law of castor oil, which behaves very similar to biological soft tissue, is approximately a power law with exponent $\gamma = 1.66$ and $\hat{\alpha}_0 \approx 4 \cdot 10^{-2} \frac{1}{cm\,(MHz)^\gamma}$, i.e.

$$\alpha_{pl}(\omega) \approx 4 \cdot 10^{-2} \cdot \omega^{1.66} \cdot cm^{-1} \qquad (\omega \text{ in MHz}).$$

The sound speed of castor oil is $1490 \cdot \frac{m}{s}$ at 25 degree Celsius. In units of cm and MHz we have

$$c_0 \approx 0.15 \cdot cm\,MHz \,.$$

Since (4.28) approximates the power law (cf. Fig. 4.4) it follows that

$$\hat{\alpha}_0 \, |\omega|^\gamma \approx \frac{\alpha_0 \, \sin(\frac{\pi}{2}(\gamma - 1))}{2\,c_0\,\tau_0} \, |\tau_0\,\omega|^\gamma$$

and consequently the coefficients of (4.28) satisfy

$$\alpha_0 \approx \frac{2\,c_0\,\hat{\alpha}_0}{\tau_0^{(\gamma - 1)} \, \sin(\frac{\pi}{2}(\gamma - 1))} \approx 6 \,.$$

We note that the relaxation time is $\tau \approx 10^{-4}\,MHz^{-1}$ for liquids (cf. [21]).

For the calculation of the singular value decomposition of the discretized kernel of the integral equation (4.100) we used a frequency range $\omega \in [-80, 80]$MHz and step size $\Delta\omega = \frac{2\pi}{N-1}$ MHz with $N = 2^9$. The time interval has been set to $[0, \frac{2\pi}{\Delta\omega}]$MHz^{-1} and a step size $\Delta t = \frac{2\pi}{80}$ MHz^{-1} was used.

The upper left picture in Fig. 4.5 visualizes the discretized kernel of the integral equation (4.100) for $R_0 = R$, i.e., when $\mathbf{x}_0$ is directly on the surface of the object of interest. The upper right picture shows the singular values of the discretized kernel in a logarithmic scale. Two properties of the singular values become apparent:

1. For large indices the decay rate is exponential, which can be seen from the linear decay in the logarithmic scale.
2. Secondly, there is a range of indices, where the singular values do not decay that rapidly. As a consequence, for solving the integral equation this means that the Fourier coefficients of p_0 according to the first block of singular values can be determined in a stable manner.

For increasing distance $L = |\mathbf{x}_0| - R$ of $\mathbf{x}_0$ to the object the singular values of the discretization of the integral equation (4.100) show a drastically more exponentially decay rate for increasing L (see bottom right picture Fig. 4.5). This means that if the object is further away from $\mathbf{x}_0$ attenuation is more drastically, and solution of the integral equation is more unstable. We analyze the dependence of the number of largest singular values from L. For this purpose we denote by n_{cut} the index of the singular value that is about 0.1% of the maximal singular value. For the numerical solution of (4.100) it means that if we make a truncated singular valued decomposition with only n_{cut} singular values, the error amplification can be bounded by a factor 1000. The dependence of n_{cut} on L is shown in the lower left picture of Fig. 4.5. The picture reveals that for increasing distance (from about $2cm$) only about four Fourier modes of p_0 are significant when a maximal error amplification of a factor 1000 is required. This reveals that in general the solution of the integral equation (4.100) is significantly ill–posed and worse if the data recording is far away from the object.

Example 3. An analogous numerical example as in Example 2 for the case $\gamma = 1.1$ is presented in Fig. 4.6. From the lower left picture of Fig. 4.6, we see that if the distance is about $2 \cdot cm$ from the boundary of the object, then 17 singular values are available for the numerical estimation.

Example 4. An analogous numerical example as in Example 2 for the case $\gamma = 2$ is presented in Fig. 4.7. From the lower left picture of Fig. 4.7, we see that if the distance is about $2 \cdot$ cm from the boundary of the object, then only 4 singular values are available for the numerical estimation.

Example 5. An analogous numerical example as in Example 2 for the frequency power law

$$\alpha_*(\omega) = \alpha_0^{pl} \cdot (-i\,\omega)^{0.66} \qquad (\alpha_0^{pl} \text{ as in Example 2})$$

is presented in Fig. 4.8. From the lower left picture of this figure, we see that if the distance is about $2 \cdot$ cm from the boundary of the object, then 77 singular values are available for the numerical estimation. If the distance is about $4cm$, then 46 singular values are available.

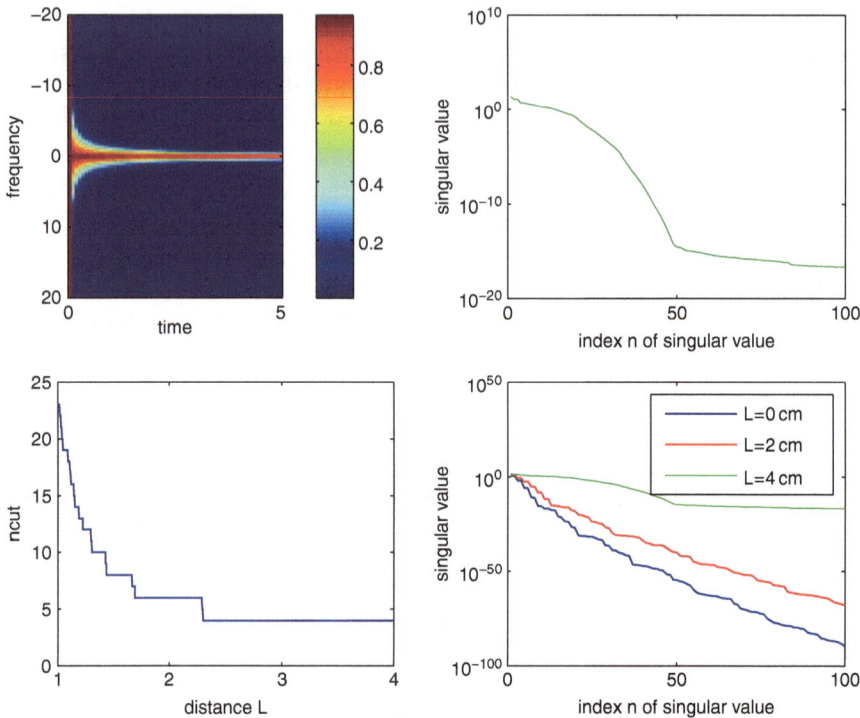

Fig. 4.5. Case: $\gamma = 1.66$ (castor oil). The upper left and right pictures visualize the kernel $\mathcal{F}\{\mathcal{M}\}(\omega, t')$ and its singular values for $L := R_0 - R = 0cm$. The lower right and left pictures visualize $\mathcal{F}\{\mathcal{M}\}(\omega, t')$ for the detector distances $L = 0 \cdot cm$, $L = 2 \cdot cm$ and $L = 4 \cdot cm$ and the respective indices n_{cut} for which the singular values are about 0.1 per cent of the maximal singular value

Comparing all numerical examples shows that the larger γ (stronger attenuation), the more rapidly decrease the singular values.

4.8 Appendix: Nomenclature and Elementary Facts

Sets: B_R denotes the open ball with center at $\mathbf{0}$ and radius R. $S^n \subseteq \mathbb{R}^n$ denotes the n-dimensional unit sphere.

Real and Complex Numbers: $\mathbb{C}$ denotes the space of complex numbers, $\mathbb{R}$ the space of reals. For a complex number $c = a + \mathrm{i}\,b$ $a = \mathrm{Re}(c)$, $b = \mathrm{Im}(c)$ denote the real and imaginary parts, respectively.

For a complex number c we denote by $|c|$ the absolute value and by $\phi \in (-\pi, \pi]$ the argument. That is

$$c = |c| \exp(i\phi) \ .$$

As a consequence, when $w = r \exp(i\,\phi)$ then

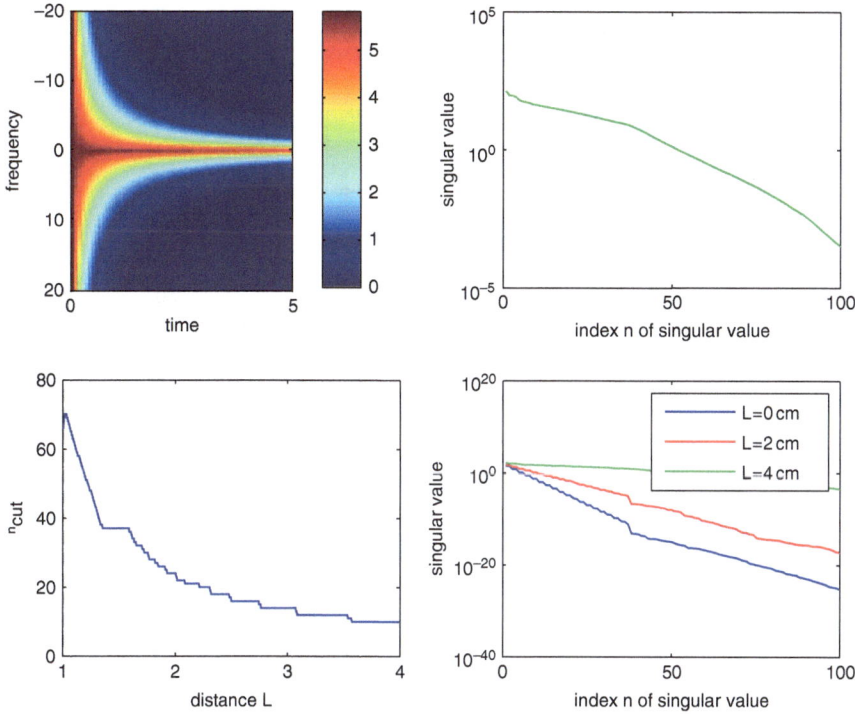

Fig. 4.6. Case: $\gamma = 1.1$. The upper left and right pictures visualize the kernel $\mathcal{F}\{\mathcal{M}\}(\omega, t')$ and its singular values. The lower right and left pictures visualize $\mathcal{F}\{\mathcal{M}\}(\omega, t')$ for the detector distances $L = 0 \cdot$ cm, $L = 2 \cdot$ cm and $L = 4 \cdot$ cm and the respective indices n_{cut} for which the singular values are about 0.1 per cent of the maximal singular value

$$ w^{\gamma} = \exp\left(\gamma \left(\log(r) + \mathrm{i}\,\phi\right)\right) . \tag{4.101} $$

Consequently w^{γ} has absolute value r^{γ} and the argument is $\gamma\phi$ modulo 2π. In this chapter all power functions are defined on $\mathbb{C}\backslash\mathbb{R}_-$. We note that

$$ w \in \mathbb{C}\backslash\mathbb{R}_- \quad \text{and} \quad \mathrm{Re}(\sqrt{w}) > 0 \quad \Rightarrow \quad \mathrm{Im}(w)\,\mathrm{Im}(\sqrt{w}) \geq 0 . \tag{4.102} $$

Differential Operators: ∇ denotes the gradient. $\nabla\cdot$ denotes divergence, and ∇^2 denotes the Laplacian.

Product: When we write $\cdot$ between two functions, then it means a pointwise product, it can be a scaler product or if the functions are vector valued an inner product. The product between a function and a number is not explicitly stated.

Composition: The composition of operators $\mathcal{A}$ and $\mathcal{B}$ is written as $\mathcal{A}\mathcal{B}$.

Special functions:

- The *signum* function is defined by

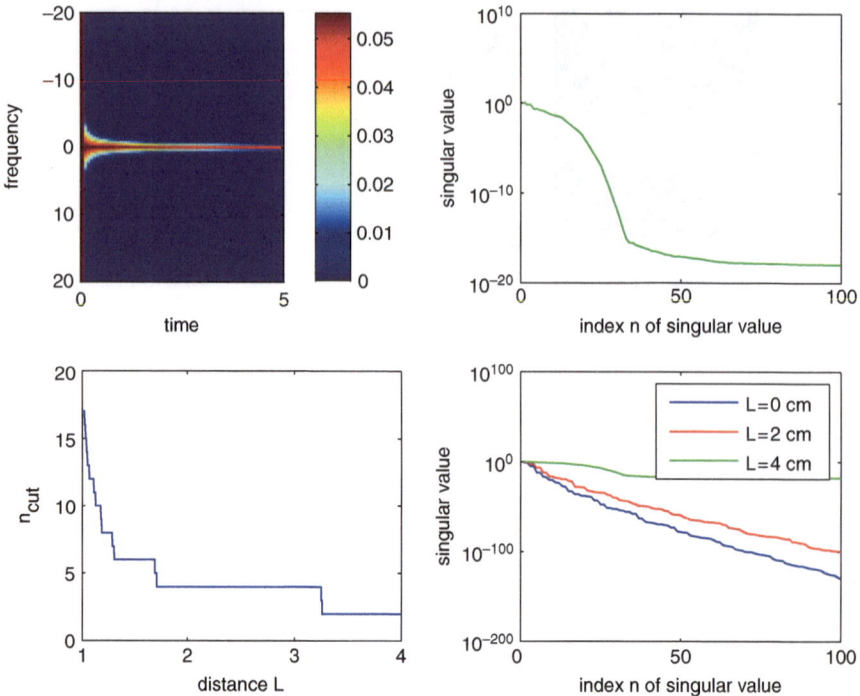

Fig. 4.7. Case: $\gamma = 1.1$. The upper left and right pictures visualize the kernel $\mathcal{F}\{\mathcal{M}\}(\omega, t')$ and its singular values. The lower right and left pictures visualize $\mathcal{F}\{\mathcal{M}\}(\omega, t')$ for the detector distances $L = 0 \cdot$ cm, $L = 2 \cdot$ cm and $L = 4 \cdot$ cm and the respective indices n_{cut} for which the singular values are about 0.1 per cent of the maximal singular value

$$\mathrm{sgn} := \mathrm{sgn}(\mathbf{x}) := \frac{\mathbf{x}}{|\mathbf{x}|} \, .$$

In $\mathbb{R}^3$ it satisfies

$$\nabla \cdot \mathrm{sgn} = \frac{2}{|\mathbf{x}|} \, . \tag{4.103}$$

- The *Heaviside* function

$$H := H(t) := \begin{cases} 0 & \text{for } t < 0 \\ 1 & \text{for } t > 0 \end{cases}$$

satisfies

$$H := \frac{1}{2}(1 + \mathrm{sgn}) \, .$$

- The δ-distribution is the derivative of the Heaviside function at 0 and is denoted by $\delta_t := \delta_t(t)$. In our terminology δ_t denotes a *one*-dimensional distribution. Sometimes, if the context is clear, we will omit the subscript at the δ-distributions.

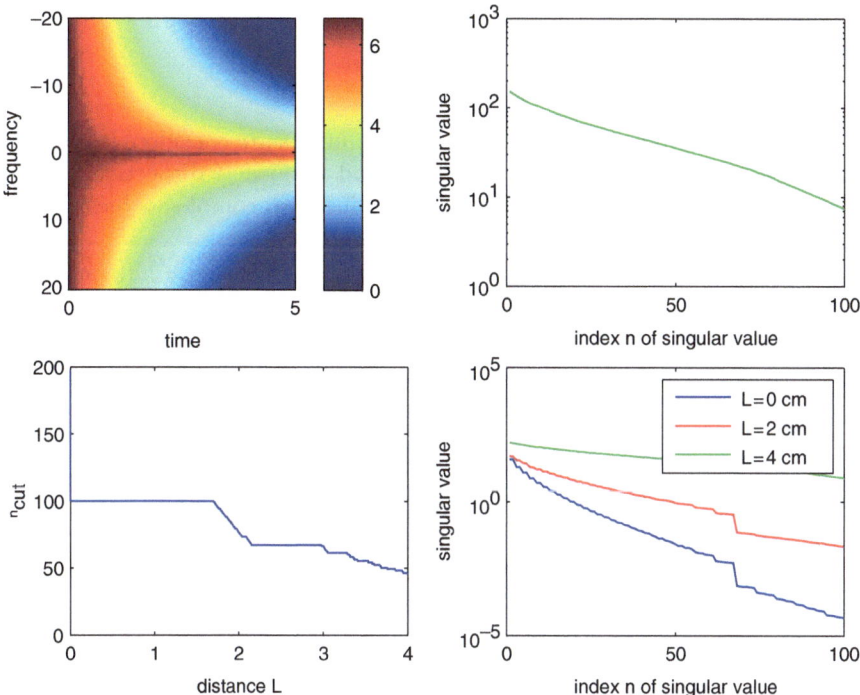

Fig. 4.8. Case: $\gamma = 0.66$. The upper left and right pictures visualize the kernel $\mathcal{F}\{\mathcal{M}\}(\omega, t')$ and its singular values. The lower right and left pictures visualize $\mathcal{F}\{\mathcal{M}\}(\omega, t')$ for the detector distances $L = 0 \cdot$ cm, $L = 2 \cdot$ cm and $L = 4 \cdot$ cm and the respective indices n_{cut} for which the singular values are about 0.1 per cent of the maximal singular value

- The three dimensional δ-distribution $\delta_{\mathbf{x}}$ is the tensor product of the three one-dimensional distributions δ_{x_i}, $i = 1, 2, 3$. Moreover,

$$\delta_{\mathbf{x},t} := \delta_{\mathbf{x},t}(\mathbf{x}, t) = \delta_{\mathbf{x}} \cdot \delta_t, \tag{4.104}$$

  is a four dimensional distribution in space and time. If we do not add a subscript δ denotes a one-dimensional δ-distribution.
- χ_Ω denotes the characteristic set of Ω, i.e., it attains the value 1 in Ω and is zero else.

Properties related to functions: $\text{supp}(g)$ denote the *support* of the function g, that is the closure of the set of points, where g does not vanish.

Derivative with respect to radial components: We use the notation

$$r := r(\mathbf{x}) = |\mathbf{x}|,$$

and denote the derivative of a function f, which is only dependent on the radial component $|\mathbf{x}|$, with respect to r (i.e., with respect to $|\mathbf{x}|$) by $\cdot'$.

Let $\beta = \beta(r)$, then it is also identified with the function $\beta = \beta(|\mathbf{x}|)$ and therefore

$$\nabla \beta = \frac{\mathbf{x}}{|\mathbf{x}|} \beta' .$$

Convolutions: Three different types of convolutions are considered: $*_t$ and $*_\omega$ denote *convolutions* with respect to time and frequency, respectively. Let f, $\hat{f}$, g and $\hat{g}$ be functions defined on the real line with complex values. Then

$$f *_t g := \int_{\mathbb{R}} f(t - t')g(t')dt', \qquad \hat{f} *_\omega \hat{g} := \int_{\mathbb{R}} \hat{f}(\omega - \omega')\hat{g}(\omega')d\omega'.$$

$*_{\mathbf{x},t}$ denotes space–time convolution and is defined as follows: Let f, g be functions defined on the Euclidean space $\mathbb{R}^3$ with complex values, then

$$f *_{\mathbf{x},t} g := \int_{\mathbb{R}^3} \int_{\mathbb{R}} f(\mathbf{x} - \mathbf{x}', t - t')g(\mathbf{x}', t')d\mathbf{x}'dt' .$$

Fourier transform: For more background we refer to [17,28,33,42,54]. All along this chapter $\mathcal{F}\{\cdot\}$ denotes the *Fourier transformation* with respect to t, and the *inverse Fourier transform* $\mathcal{F}^{-1}\{\cdot\}$ is with respect to ω. In this chapter we use the following definitions of the transforms:

$$\mathcal{F}\{f\}(\omega) = \frac{1}{\sqrt{2\pi}} \int_{\mathbb{R}} \exp(i\omega t) f(t)dt ,$$

$$\mathcal{F}^{-1}\{\hat{f}\}(t) = \frac{1}{\sqrt{2\pi}} \int_{\mathbb{R}} \exp(-i\omega t) \hat{f}(\omega)d\omega .$$

The Fourier transform and its inverse have the following properties:

1.

$$\mathcal{F}\left\{\frac{\partial}{\partial t} f\right\}(\omega) = (-i\omega)\mathcal{F}\{f\}(\omega) .$$

2.

$$\mathcal{F}\{f \cdot g\} = \frac{1}{\sqrt{2\pi}} \mathcal{F}\{f\} *_\omega \mathcal{F}\{g\} \text{ and}$$

$$\mathcal{F}\{f\} \cdot \mathcal{F}\{g\} = \frac{1}{\sqrt{2\pi}} \mathcal{F}\{f *_t g\} ,$$

$$\mathcal{F}^{-1}\{\hat{f} \cdot \hat{g}\} = \frac{1}{\sqrt{2\pi}} \mathcal{F}^{-1}\{\hat{f}\} *_t \mathcal{F}^{-1}\{\hat{g}\} \text{ and}$$

$$\mathcal{F}^{-1}\{\hat{f}\} \cdot \mathcal{F}^{-1}\{\hat{g}\} = \frac{1}{\sqrt{2\pi}} \mathcal{F}^{-1}\{\hat{f} *_\omega \hat{g}\} .$$

3. For $a \in \mathbb{R}$

$$\mathcal{F}\{f(t - a)\}(\omega) = \exp(ia\omega) \cdot \mathcal{F}\{f(t)\}(\omega)$$

4. The δ-distribution at $a \in \mathbb{R}$ satisfies

$$\delta_t(t - a) = \frac{1}{\sqrt{2\pi}} \mathcal{F}^{-1}\{\exp(i a\omega)\}(t) .$$

5. Let f be real and even, odd respectively, then $\mathcal{F}\{f\}$ is real and even, imaginary and odd, respectively.
6. The Fourier transformation of a tempered distribution is a tempered distribution.
7. Let $\tau_1, \tau_2 \in \mathbb{R}$. If f_1 and f_2 are two distributions with support in $[\tau_1, \infty)$ and $[\tau_2, \infty)$, respectively, then $f_1 * f_2$ is well-defined and (cf. [17])

$$\text{supp}(f_1 * f_2) \subseteq \text{supp}(f_1) + \text{supp}(f_2) \subseteq [\tau_1 + \tau_2, \infty). \tag{4.105}$$

The Hilbert transform for L^2-functions is defined by

$$\mathcal{H}\{f\}(t) = \frac{1}{\pi} \fint_{\mathbb{R}} \frac{f(s)}{t-s} ds \,,$$

where $\fint_{\mathbb{R}} f(s)ds$ denotes the Cauchy principal value of $\int_{\mathbb{R}} f(s)ds$.

A more general definition of the Hilbert transform can be found in [2]. The Hilbert transform satisfies

- $\mathcal{H}\{\mathcal{F}\{f\}\}(\omega) = -i\mathcal{F}\{\text{sgn}f\}(\omega)$,
- $\mathcal{H}\{\mathcal{H}\{f\}\} = -f$.

From the first of these properties the Kramers-Kronig relation can be formally derived as follows. Since $f(t)$ is a causal function if and only if $f = H \cdot f$ and $H = (1 + \text{sgn})/2$, it follows that $\mathcal{F}\{f\} = [\mathcal{F}\{f\} + i\mathcal{H}\{\mathcal{F}\{f\}\}]/2$, which is equivalent to $\mathcal{F}\{f\} = i\mathcal{H}\{\mathcal{F}\{f\}\}$, i.e.

$$\text{Re}(\mathcal{F}\{f\}) = -\text{Im}(\mathcal{H}\{\mathcal{F}\{f\}\}) \qquad \text{and} \qquad \text{Im}(\mathcal{F}\{f\}) = \text{Re}(\mathcal{H}\{\mathcal{F}\{f\}\}).$$

The inverse Laplace transform of f is defined by

$$\mathcal{L}^{-1}\{f\}(t) = \begin{cases} 0 & \text{for } t < 0, \\ \frac{1}{2\pi i} \int_{\gamma - i\infty}^{\gamma + i\infty} \exp(st) f(s)ds, & \text{for } t > 0, \end{cases}$$

where γ is appropriately chosen.

The inverse Laplace transform satisfies (see e.g. [16])

$$\mathcal{L}^{-1}\{h(s-a)\}(t) = \exp(at)\,\mathcal{L}^{-1}\{h(s)\}(t) \quad \text{for all } a, t \in \mathbb{R} \tag{4.106}$$

and

$$\mathcal{L}^{-1}\{s^{-r}\}(t) = \frac{H(t)t^{r-1}}{\Gamma(r)} \qquad (r > 0). \tag{4.107}$$

Acknowledgements

This work has been supported by the Austrian Science Fund (FWF) within the national research network Photoacoustic Imaging in Biology and Medicine, project S10505-N20.

References

1. M. Agranovsky, P. Kuchment, Uniqueness of reconstruction and an inversion procedure for thermoacoustic and photoacoustic tomography with variable sound speed, Inv. Probl. **23**(5), 2089–2102 (2007)
2. E.J. Beltrami, M.R. Wohlers, *Distributions and the Boundary Values of Analytic Functions*. Academic Press, New York and London (1966)
3. P. Burgholzer, J. Bauer-Marschallinger, H. Grün, M. Haltmeier, G. Paltauf, Temporal back-projection algorithms for photoacoustic tomography with integrating line detectors. Inverse Probl. **23**(6), 65–80 (2007)
4. P. Burgholzer, H. Grün, M. Haltmeier, R. Nuster, G. Paltauf, in *Compensation of acoustic attenuation for high-resolution photoacoustic imaging with line detectors*. ed. by A.A. Oraevsky, L.V. Wang, Photons Plus Ultrasound: Imaging and Sensing 2007: The Eighth Conference on Biomedical Thermoacoustics, Optoacoustics, and Acousto-optics. vol. 6437 Proceedings of SPIE, p. 643724. SPIE (2007)
5. P. Burgholzer, H. Roitner, J. Bauer-Marschallinger, G. Paltauf, Image Reconstruction in Photoacoustic Tomography Using Integrating Detectors Accounting for Frequency-Dependent Attenuation. vol. 7564 Proc. SPIE p. 75640O. (2010)
6. W. Chen, S. Holm, Fractional Laplacian time-space models for linear and nonlinear lossy media exhibiting arbitrary frequency power-law dependency. J. Acoust. Soc. Am. **115**(4), 1424–1430 (2004)
7. W.F. Cheong, S.A. Prahl, A.J. Welch, A review of the optical properties of biological tissues. IEEE J. Quantum Electron **26**(12), 2166–2185 (1990)
8. R. Dautray, J.-L. Lions, Mathematical Analysis and Numerical Methods for Science and Technology. Vol. 1, Springer, New York (2000)
9. R. Dautray, J.-L. Lions, Mathematical Analysis and Numerical Methods for Science and Technology. Vol. 2, Springer, New York (2000)
10. R. Dautray, J.-L. Lions, Mathematical Analysis and Numerical Methods for Science and Technology. Vol. 5, Springer, New York (2000)
11. D. Finch, S. Patch, Rakesh, Determining a function from its mean values over a family of spheres. Siam J. Math. Anal. **35**(5), 1213-1240 (2004)
12. C. Gasquet, P. Witomski, Fourier Analysis and Applications. Springer, New York (1999)
13. V.E. Gusev, A.A. Karabutov, Laser Optoacoustics. American Institute of Physics, New York (1993)
14. M. Haltmeier, O. Scherzer, P. Burgholzer, P. Paltauf, Thermoacoustic imaging with large planar receivers. Inverse Probl. **20**(5), 1663-1673 (2004)
15. A. Hanyga, M. Seredynska, Power-law attenuation in acoustic and isotropic anelastic media. Geophys. J. Int. **155**, 830-838 (2003)
16. H. Heuser, Gewöhnliche Differentialgleichungen. 2nd edn. Teubner, Stuttgart (1991)
17. L. Hörmander, The Analysis of Linear Partial Differential Operators I. 2nd edn. Springer, New York (2003)
18. Y. Hristova, P. Kuchment, L. Nguyen, Reconstruction and time reversal in thermoacoustic tomography in acoustically homogeneous and inhomogeneous media. Inverse Probl. **24**(5), 055006 (25pp) (2008)
19. F. John, Partial Differential Equations. Springer, New York (1982)

20. A.A. Kilbas, H.M. Srivastava, J.J. Trujillo, in *Theory and Applications of Fractional Differential Equations*, North-Holland Mathematics Studies. vol. 204 (Elsevier Science B.V., Amsterdam 2006)

21. L.E. Kinsler, A.R. Frey, A.B. Coppens, J.V. Sanders, Fundamentals of Acoustics. Wiley, New York (2000)

22. R. Kowar, O. Scherzer, X. Bonnefond, Causality analysis of frequency-dependent wave attenuation. Math. Meth. Appl. Sci., **22**, 108–124 (2011). DOI: 10.1002/mma.1344

23. R.A. and Kiser, W. L. and Miller, K. D. and Reynolds, H. E.: Thermoacoustic CT: imaging principles. Proc. SPIE, vol. 3916, pp. 150–159 (2000)

24. G. Ku, X. Wang, G. Stoica, L.V. Wang, Multiple-bandwidth photoacoustic tomography. Phys. Med. Biol. **49**, 1329–1338 (2004)

25. L.A. Kunyansky, Explicit inversion formulae for the spherical mean Radon transform. Inverse Probl. **23**, 373–383 (2007)

26. P. Kuchment, L.A. Kunyansky, Mathematics of thermoacoustic and photoacoustic tomography. Eur. J. Appl. Math. **19**, 191–224 (2008)

27. L.D. Landau, E.M. Lifschitz, Lehrbuch der theoretischen Physik, Band VII: Elastizitätstheorie. Akademie, Berlin (1991)

28. M.J. Lighthill, Introduction to Fourier Analysis and Generalized Functions. Student's Edition. Cambridge University Press, London (1964)

29. A.I. Nachman, J.F. Smith III, R.C. Waag, An equation for acoustic propagation in inhomogeneous media with relaxation losses. J. Acoust. Soc. Am. **88**(3) 1584–1595 (1990)

30. Oraevsky, A. and Wang, L.V., editors: *Photons Plus Ultrasound: Imaging and Sensing 2007: The Eighth Conference on Biomedical Thermoacoustics, Optoacoustics, and Acousto-optics*, (SPIE Publishing, Bellingham, WA, 2007), Vol. 6437, p. 82

31. S.K. Patch, O. Scherzer, Special section on photo- and thermoacoustic imaging. Inverse Probl. **23**, S1–S122 (2007)

32. S.K. Patch, A. Greenleaf, Equations governing waves with attenuation according to power law. *Technical report*, Department of Physics, University of Wisconsin-Milwaukee (2006)

33. A. Papoulis, *The Fourier Integral and its Applications*. McGraw-Hill, New York (1962)

34. I. Podlubny, in *Fractional Differential Equations*, Mathematics in Science and Engineering. vol. 198 (Academic Press Inc., San Diego, CA 1999)

35. D. Razansky, M. Distel, C. Vinegoni, R. Ma, N. Perrimon, R.W. Köster, V. Ntziachristos, Multispectral opto-acoustic tomography of deep-seated fluorescent proteins in vivo. Nat. Photonics **3**, 412-417 (2009)

36. P.J. La Riviére, J. Zhang, M.A. Anastasio, Image reconstruction in optoacoustic tomography for dispersive acoustic media. Opt. Lett. **31**(6), 781–783 (2006)

37. O. Scherzer, H. Grossauer, F. Lenzen, M. Grasmair, M. Haltmeier, Variational Methods in Inmaging. Springer, New York (2009)

38. N.V. Sushilov, R.S.C. Cobbold, Frequency-domain wave equation and its time-domain solution in attenuating media. J. Acoust. Soc. Am. **115**, 1431–1436 (2005)

39. T.L. Szabo, Time domain wave equations for lossy media obeying a frequency power law. J. Acoust. Soc. Amer. **96**, 491–500 (1994)

40. T.L. Szabo, Causal theories and data for acoustic attenuation obeying a frequency power law. J. Acoust. Soc. Amer. **97**, 14–24 (1995)

41. A.C. Tam, Applications of photoacoustic sensing techniques. Rev. Mod. Phys. **58**(2), 381–431 (1986)
42. E.C. Titchmarch, *Theory of Fourier Integrals*. Clarendon Press, Oxford (1948)
43. K.R. Waters, M.S. Hughes, G.H. Brandenburger, J.G. Miller, On a time-domain representation of the Kramers-Krönig dispersion relation. J. Acoust. Soc. Amer. **108**(5), 2114–2119 (2000)
44. K.R. Waters, J. Mobely, J.G. Miller, Causality-Imposed (Kramers-Krönig) Relationships Between Attenuation and Dispersion. IEEE Trans. Ultrason. Ferroelect. Freq. Contr. **52**(5) 822–833 (2005)
45. Webb, S., ed. *The Physics of Medical Imaging*. Institute of Physics Publishing, Bristol, Philadelphia (2000); reprint of the 1988 edition.
46. L.V. Wang, Prospects of photoacoustic tomography. Med. Phys. **35**(12), 5758–5767 (2008)
47. X.D. Wang, Y.J. Pang, G. Ku, X.Y. Xie, G. Stoica, L.V. Wang, Noninvasive Laser-Induced Photoacoustic Tomography for Structural and Functional *in Vivo* Imaging of the Brain Nat. Biotechnol. **21**(7), 803–806 (2003)
48. Y. Xu, D. Feng, L.V. Wang, Exact Frequency-Domain Reconstruction for Thermoacoustic Tomography – I: Planar Geometry. IEEE Trans. Med. Imag. **21**(7) 823–828 (2002)
49. Y. Xu, M. Xu, L.V. Wang, Exact Frequency-Domain Reconstruction for Thermoacoustic Tomography – II: Cylindrical Geometry. IEEE Trans. Med. Imag. **21**(7), 823–828 (2002)
50. M. Xu, Y. Xu, L.V. Wang, Time-Domain Reconstruction Algorithms and Numerical Simulation for Thermoacoustic Tomography in Various Geometries. IEEE Trans. Biomed. Eng. **50**(9), 1086–1099 (2003)
51. Y. Xu, L.V. Wang, G. Ambartsoumian, P. Kuchment, Reconstructions in limited-view thermoacoustic tomography. Med. Phys. **31**(4), 724–733 (2004)
52. Xu, M. and Wang, L. V.: Universal back-projection algorithm for photoacoustic computed tomography. Phys. Rev. E 71 (2005) [7 pages] Article ID 016706.
53. M. Xu, L.V. Wang, Photoacoustic imaging in biomedicine. Rev. Sci. Instrum. **77**(4), 1–22 (2006) Article ID 041101.
54. Yosida, K.: Functional analysis. 5th edn., Springer, New York (1995)
55. Zhang, E.Z. and Laufer, J. and Beard, P.: Three-dimensional photoacoustic imaging of vascular anatomy in small animals using an optical detection system. (SPIE Publishing, Bellingham, WA, 2007), Vol. 6437, p. 82
56. H. Zhang, K. Maslov, G. Stoika, V.L. Wang, Functional photoacoustic microscopy for high-resolution and noninvasive in vivo imaging. Nat. Biotechnol. **24**, 848–851 (2006)

5

Quantitative Photoacoustic Tomography

Hao Gao[1], Stanley Osher[1], and Hongkai Zhao[*,2]

[1] Department of Mathematics, University of California, Los Angeles, CA 90095, USA haog@math.ucla.edu, sjo@math.ucla.edu
[2] Department of Mathematics, University of California, Irvine, CA 92697, USA zhao@math.uci.edu

Summary. This chapter focuses on quantitative photoacoustic tomography to recover optical maps from the deposited optical energy. After a brief overview of models, theories and algorithms, we provide an algorithm for large-scale 3D reconstructions, so-called gradient-based bound-constrained split Bregman method (GBSB).

5.1 Introduction

Photoacoustic tomography (PAT), a synergistic combination of ultrasound and optical imaging, has recently emerged as a potential imaging method to resolve optical contrasts with accurate quantification and high resolution [31, 38, 39]. See also Chap. 3.

The first inverse problem in PAT concerns about the reconstruction of the deposited optical energy from the time-dependent boundary measurement of the acoustic pressure. Explicit inversion formulas exist for a large class of geometries of interest, when the problem is in free space, with constant sound speed, and without accounting of acoustic attenuation. The major efforts for this first step have been focused on the situations when any of aforementioned conditions is violated so that no explicit formulas exist. We refer the reader to other chapters in this volume and their references for the discussions of different topics in the first inverse problem in PAT.

The second inverse problem in PAT, so-called quantitative PAT (QPAT) consists of reconstructing optical maps, particularly the absorption coefficient or the chromophores, from the deposited optical energy that is recovered from the first step. In this chapter, we assume that the deposited optical energy is known and will focus on the simultaneous reconstruction of absorption coefficient and scattering coefficient. Please note that although the absorption map is usually of the major clinic interest, it is necessary to consider the scattering map in reconstruction as well in order to accurately reconstruct the absorption map when the scattering coefficient is unknown.

H. Ammari (ed.), *Mathematical Modeling in Biomedical Imaging II*,
Lecture Notes in Mathematics 2035, DOI 10.1007/978-3-642-22990-9_5,
© Springer-Verlag Berlin Heidelberg 2012

There are mainly three methodologies for QPAT. First, using single optical illumination and assuming the scattering map is known, we can recover the absorption map [6, 15, 40]; second, using multi-wavelength illuminations and assuming the spectral model of optical coefficients, both absorption and scattering maps can be obtained [14]; third, using multiple optical illumination, so-called multiple-source QPAT (MS-QPAT), both absorption and scattering maps can be recovered [9, 26], for which the reconstruction uniqueness and stability estimates can be rigorously established [BU10]. MS-QPAT was also considered in [36, 41] for the reconstruction of the absorption map. Using an asymptotic approach in the spirit of Chap. 1, efficient methods for QPAT in the context of small-volume absorbers are derived in [3]. We will mainly focus on MS-QPAT in this chapter.

The major contribution of this chapter is to provide an algorithm for reconstructing optical maps in large-scale 3D QPAT, so-called GBSB, i.e., the gradient-based bound-constrained split Bregman method, based on the proposed method in [26]. For the enhanced stability with respect to the initial guess or the noise, the simple bounds are imposed, and the solution is regularized with total variation (TV) norm [37]. These two strategies are particularly important for scattering reconstruction, and thus for accurate absorption reconstruction. To be suitable for the large-scale computation, GBSB utilizes Quasi-Newton method as the inner loop with the computation of gradients rather than Jacobians, and split Bregman method as the outer loop that is particularly efficient for L_1-type optimization including TV-regularized problem.

The chapter is organized as follow. We first briefly overview existing models, theories, and algorithms of QPAT in Sect. 5.2; then discuss the ingredients of GBSB for the large-scale 3D QPAT, and present simulation results in Sect. 5.3; last end with a discussion section.

5.2 Overview of QPAT

5.2.1 Forward Models

The light migration in the photoacoustic imaging can be modeled by diffusion approximation (DA), a first-order phase approximation of radiative transport equation (RTE) in spherical harmonic bases [1, 19]. DA so far is the most popular since its solutions can be solved practically and the solution methods are relatively simple [1]. In contrast, although RTE is more accurate than DA, particularly in non-diffusive region, it hasn't been widely used mainly because it is challenging to solve RTE practically. However, the practical solvers of RTE on a daily laptop can be potentially available, advanced by state-of-art numerical algorithms [2, 23, 30] and novel computer architectures [32]. See Chap. 1 for a detailed discussion on light propagation modeling.

Diffusion Approximation (DA)

As shown in Chap. 1, in DA, the photon density is simplified to be dependent only on the spatial variable x. On a bounded domain Ω with smooth boundary Γ, the DA with Robin boundary condition is

$$-\nabla \cdot (D(\mathbf{x})\nabla\phi(\mathbf{x})) + \mu_a(\mathbf{x})\phi(\mathbf{x}) = q(\mathbf{x}), \ \mathbf{x} \in \Omega,$$
$$2\kappa D(\mathbf{x})(\mathbf{n} \cdot \nabla\phi(\mathbf{x})) + \phi(\mathbf{x}) = q_b(\mathbf{x}), \ \mathbf{x} \in \Gamma, \quad (5.1)$$

where μ_a is the absorption coefficient, μ_s the scattering coefficient, $\mu_s' = (1 - g)\mu_s$ the reduced scattering coefficient with the anisotropic scattering factor g (see (5.8)), D so-called diffusion coefficient that can be defined by $D = 1/(3\mu_s')$, and κ the constant for coupling the refraction index mismatch at the boundary, that can be usually determined through data fitting [1]. Please note that DA is valid only in the diffusive regime, e.g., $\mu_a << \mu_s'$ and $\mathbf{x}$ is at least a few mean free paths away from the source. We refer the readers to [19] for further discussions of DA and [1] for DA based optical tomographic problems.

After discretization, DA (5.1) can be formulated as the linear system

$$A\Phi = Q. \quad (5.2)$$

The system matrix A, the source term Q and the discretized density Φ in (5.2) are specific to discretization methods. For example, using finite element method (FEM) [28] that is popular for DA solutions on Ω with irregular shapes, the photon density in piecewise-linear bases $\{\varphi_i, i \leq N_p\}$ with N_p nodes is

$$\phi(\mathbf{x}) = \sum_i \phi_i \varphi_i(\mathbf{x}) \text{ and } [\Phi]_i = \phi_i. \quad (5.3)$$

On the other hand, we discretize the optical coefficients in piecewise constants, i.e., $\{(\mu_{a,i}, \mu_{s,i}'), i \leq N_t\}$ with N_t elements in the mesh.

With this discretization, we have

$$[A]_{ij} = \int_\Omega D(\nabla\varphi_i \cdot \nabla\varphi_j)d\mathbf{x} + \int_\Omega \mu_a\varphi_i\varphi_jd\mathbf{x} + \frac{1}{2\kappa}\int_\Gamma \varphi_i\varphi_jd\mathbf{x} \quad (5.4)$$

and

$$[Q]_j = \int_\Omega q\varphi_jd\mathbf{x} + \frac{1}{2\kappa}\int_\Gamma q_b\varphi_jd\mathbf{x}. \quad (5.5)$$

The available data in DA based QPAT have the form as the product of the absorption coefficient and the photon density, i.e., $\mu_a(\mathbf{x})\phi(\mathbf{x})$, which is assumed to be given here for all $x \in \Omega$ or can be reconstructed through acoustic inversion. In order to utilize the data in the discretized settings, we assume there are N_d numerical detectors that are uniformly distributed across the domain Ω. Although N_d can be arbitrarily large, we will let it be equal to the number of freedom of the discretized μ_a to avoid redundancy, i.e., $N_d = N_t$. Let us also assume that there are N_s optical illuminations in the setting of MS-QPAT, each of which has a flux distribution Φ_i by solving (5.2) with the formula (5.4) and (5.5). Then the discrete mapping of the data can be thought of as the following linear functional with respect to μ_a and Φ

$$F_j(\mu_a, \Phi_i) = \mu_{a,j} \sum_{j'=1}^{D} \alpha_{jj'} [\Phi_i]_{j'}, \ i \le N_s, j \le N_t, \tag{5.6}$$

where $\alpha_{jj'}$'s are the interpolation weights for the jth detector with $\sum_{j'} \alpha_{jj'} = 1$ and D, the degree of freedom of the used element (e.g., $D = 3$ for the triangular element).

Radiative Transfer Equation (RTE)

Here we briefly recall the RTE. We refer to Chap. 1 for the derivation of the scattering theory for the RTE in an inhomogeneously absorbing medium. See also [19] for a thorough treatment of RTE and [30] for its introductory numerical methods.

On a bounded domain Ω with angular space $\hat{S}$ (e.g., unit sphere in three dimensions (3D)), the steady-state RTE is

$$\hat{s} \cdot \nabla \phi(\mathbf{x}, \hat{s}) + \mu_t(\mathbf{x}) \phi(\mathbf{x}, \hat{s}) = \mu_a(\mathbf{x}) \oint_{\hat{S}} f(\hat{s}, \hat{s}') \phi(\mathbf{x}, \hat{s}') d\hat{s}' + q(\mathbf{x}, \hat{s}), \ (\mathbf{x}, \hat{s}) \in \Omega \times \hat{S}. \tag{5.7}$$

Now, the photon flux $\phi(\mathbf{x}, \hat{s})$ has not only spatial component, but also angular component, i.e., at position $\mathbf{x}$ in direction $\hat{s}$. The parameters in (5.7) are absorption coefficient μ_a, scattering coefficient μ_s, transport coefficient $\mu_t := \mu_a + \mu_s$ and the scattering kernel f. Very often f is rotationally invariant, i.e., $f(\hat{s}, \hat{s}') = f(\hat{s} \cdot \hat{s}')$, and is normalized with $\oint f(\hat{s} \cdot \hat{s}') d\hat{s}' = 1$. For example, a popular f for modeling photon transport in tissues is

$$f(\hat{s} \cdot \hat{s}') = \frac{1 - g^2}{4\pi(1 + g^2 - 2g\hat{s} \cdot \hat{s}')^{3/2}}, \tag{5.8}$$

so-called Henyey-Greenstein (H-G) function (3D version), where the anisotropy parameter $0 \le g \le 1$ measures the forward peaking of the scattering.

Given the normal $\hat{n}$ at the boundary Γ, let $\Gamma^+(\Gamma^-)$ represent $(\mathbf{x}, \hat{s}) \in \partial\Omega \times \hat{S}$ with $\hat{s} \cdot \hat{n} > 0 (\le 0)$. Let $n_i(n_o)$ be the refraction index of the medium (environment). The reflection boundary condition is

$$\phi(\mathbf{x}, \hat{s}) = R(\mathbf{x}, \hat{s}, \hat{s}') \phi(\mathbf{x}, \hat{s}') + q_b(\mathbf{x}, \hat{s}), \quad (\mathbf{x}, \hat{s}) \in \Gamma^-, \ (\mathbf{x}, \hat{s}') \in \Gamma^+, \tag{5.9}$$

where q_b is the boundary source, R is the reflection energy ratio that can be computed through Fresnel formula [29] and $\hat{s}'$ is the angle that reflects into $\hat{s}$ at the boundary $\mathbf{x}$. For the vacuum boundary condition that there is no refraction index mismatch, i.e., $n_i = n_o$, $R = 0$ in (5.9). Please note that the boundary condition for RTE is prescribed only for the incoming flux ϕ on Γ^-.

Similarly, RTE (5.7) along with the boundary condition (5.9) can be formally set up as the linear system (5.2), although (5.2) is rarely explicitly formulated for RTE due to its memory requirement. For example, following [23], we can discretize the angular variable $\hat{s}$ with FEM in piecewisely linear bases $\{\varphi_m^a, m \leq M\}$ with M angular directions and the spatial variable $\mathbf{x}$ with Discontinuous Galerkin method (DG) [17] in piecewisely linear bases $\{\varphi_{ij}^s, i \leq N_t, j \leq D\}$ (e.g., $D = 4$ on tetrahedral meshes) with N_t spatial elements. That is,

$$\phi(\mathbf{x}, \hat{s}) = \sum_{i,j,m} \phi_{ijm} \varphi_{ij}^s(\mathbf{x}) \varphi_m^a(\hat{s}) \text{ and } [\Phi]_{ijm} = \phi_{ijm}. \tag{5.10}$$

With this discretization, we have

$$
\begin{aligned}
[A]_{ijm,i'j'm'} &= -\delta_{mm'}\delta_{ii'} \int_{\Delta_{i'}} \varphi_{ij}^s(\hat{s}_{m'} \cdot \nabla \varphi_{i'j'}^s) d\mathbf{x} \\
&+ \delta_{mm'}\delta_{ii'} \int_{\Gamma_{i'}^+} \varphi_{ij}^s \varphi_{i'j'}^s |\hat{s}_{m'} \cdot \hat{n}| d\mathbf{x} \\
&- \delta_{mm'}\delta_{T(i)i'} \int_{\Gamma_{i'}^-} \varphi_{ij}^s \varphi_{ij'}^s |\hat{s}_{m'} \cdot \hat{n}| d\mathbf{x} \\
&- \delta_{R(m)m'}\delta_{ii'} \int_{\Gamma^-} r_{mm'} \varphi_{ij}^s \varphi_{i'j'}^s |\hat{s}_{m'} \cdot \hat{n}| d\mathbf{x} \\
&+ \delta_{mm'}\delta_{ii'}\mu_{t,i} \int_{\Delta_{i'}} \varphi_{ij}^s \varphi_{i'j'}^s d\mathbf{x} + \delta_{ii'}\mu_{s,i} \int_{\Delta_{i'}} w_{mm'}\varphi_{ij}^s \varphi_{i'j'}^s d\mathbf{x}
\end{aligned}
\tag{5.11}
$$

and

$$[Q]_{i'j'm'} = \int_{\Delta_{i'}} q\varphi_{i'j'}^s d\mathbf{x} + \int_{\Gamma^-} q_b \varphi_{i'j'}^s |\hat{s}_{m'} \cdot \hat{n}| d\mathbf{x}. \tag{5.12}$$

In (5.11), Δ_i denotes the spatial element, Γ_i^+ (resp. Γ_i^-) the internal edge or surface (excluding the domain boundary Γ) with the normal direction $\mathbf{n}$ satisfying $\hat{s}_m \cdot \mathbf{n} > 0$ (resp. $\hat{s}_m \cdot \mathbf{n} < 0$), $T(i)$ the upwind flux function that localizes the element that provides upwind fluxes to the element i, $R(m)$ the reflection function that localizes the direction from which the direction m is reflected, $r_{mm'}$ the reflection weight into m from the direction m' that is discretized from R in (5.9), the optical coefficients are assumed to be piecewise constant, i.e., $\mu_{a,i}$ and $\mu_{t,i}$, the piecewisely linear scattering function $\tilde{f}$, i.e., $\tilde{f} = \sum_m f_m \varphi_m^a$, and the angularly scattering weights $w_{mm'}$, i.e., $w_{mm'} = \int_{\hat{s}} \tilde{f} \varphi_{m'}^a d\hat{s}$. We refer the readers to [23] for details of solution algorithms and [24] for error estimates.

Now, the available data in RTE based QPAT have the form as the product of the absorption coefficient and angularly-averaged fluxes, i.e.,

$\mu_a(\mathbf{x}) \int_{\hat{S}} \phi(\mathbf{x}, \hat{s}) d\hat{s}$. Same as before, let us also assume that there are N_s optical illuminations in the setting of MS-QPAT, each of which has a flux distribution Φ_i by solving (5.2) with the formula (5.11) and (5.12). Then the discrete data for RTE can be formulated as follow

$$F_j(\mu_a, \Phi_i) = \mu_{a,j} \sum_{j'=1}^{D} \alpha_{jj'} \left(\sum_{m'=1}^{M} w_{m'} [\Phi_i]_{jj'm'} \right), \ i \leq N_s, j \leq N_t, \quad (5.13)$$

where $\alpha_{jj'}$'s again are the interpolation weights from the jth detector with $\sum_{j'} \alpha_{jj'} = 1$, and the angular weight $w_m = \int_{\hat{S}} \varphi_m^a d\hat{s}$.

5.2.2 Theory

In this section, we will discuss some established uniqueness and stability estimates mainly for MS-QPAT under single optical spectrum.

For QPAT based on DA, the reconstruction uniqueness and stability estimates were established for cases with at least two optical illuminations [12]. A non-uniqueness result was obtained in [10] for the case with single optical illumination, in which the available data are assumed to have the form $\mu_a \phi^2$ rather than $\mu_a \phi$ that comes from a different acousto-optic model [11]. Assuming the available data have the form $\gamma \mu_a \phi$ (γ: the Grüneisen coefficient), it was shown in [9] that only two out of the three coefficients (μ_a, D, γ) can be uniquely recovered even for an arbitrary number of illuminations.

The QPAT based on RTE was analyzed in [7]. The analysis assumes that the measurement operator can be decomposed into singular components, i.e., the ballistic component, the single scattering component and the multiple scattering component. With the ballistic component (the most singular component), μ_a and μ_s in (5.7) can be reconstructed; with the single scattering component, the anisotropy coefficient g in H-G function (5.8) can be uniquely reconstructed. All the above parameters are obtained with Hölder-type stability. We refer the interested readers to [7] for further details. Please note that the multi-source setting was not considered in [7], since it is sufficient to recover (μ_a, μ_s, g) assuming that each component can be well separated from the data. This assumption is appropriate when RTE is in the transport regime, which may limit its practical interest since in rare cases the regions of practical interest can be fully characterized by transport regime. That is RTE based QPAT with single optical illumination may not reconstruct (μ_a, μ_s, g) due to the potential failure of extracting singular components in many practical cases. However, in this case, we should still be able to reconstruct at least (μ_a, μ_s) in the setting of MS-QPAT assuming g is known, since (μ_a, μ_s') can be reconstructed for DA based MS-QPAT while the behavior of RTE is close to DA when the regime transits from being transport-like to diffusive-like. On the other hand, it is unclear whether g can be recovered as well even in MS-QPAT in the hybrid regime of transport and diffusion.

DA Based QPAT

The discussions here follow the results in [12]. Slightly different from the aforementioned DA model (5.1), the considered model is the following DA with Dirichlet boundary condition on a bounded domain X with the boundary Γ

$$-\nabla \cdot (D(\mathbf{x})\nabla\phi_i(\mathbf{x})) + \mu_a(\mathbf{x})\phi_i(\mathbf{x}) = 0, \ \mathbf{x} \in \Omega,$$
$$\phi_i(\mathbf{x}) = q_i(\mathbf{x}), \ \mathbf{x} \in \Gamma, \tag{5.14}$$

and the available data with N_s optical illuminations are

$$F := \{F_i = \mu_a(\mathbf{x})\phi_i(\mathbf{x}), \ i \leq N_s\}. \tag{5.15}$$

The object here is to reconstruct (μ_a, D) from F, that is equivalent to reconstruct (μ_a, μ'_s) . Suppose that the optical coefficients $(\mu_a, D) \in V$ are sufficiently smooth, i.e.,

$$V = \{(\mu_a, D) \text{ such that } (\mu_a, \sqrt{D}) \in C^{k+1} \times Y, \|\mu_a\|_{C^{k+1}} + \|\sqrt{D}\|_Y \leq M\}, \tag{5.16}$$

where $k \geq 1$ and Y is some subspace of C^{k+2}, the following uniqueness theorem (Theorem 1) with $N_s = 2$ and the stability estimate (Theorem 2) with $N_s = 2n$ (n: the dimension of the spatial domain X) can be obtained.

Theorem 5.1. *Let $N_s = 2$ and Ω be an open, bounded domain with C^2 boundary Γ. Assume $(\mu_a, D), (\tilde{\mu}_a, \tilde{D}) \in V$ and $D = \tilde{D}$ on Γ. Let F and $\tilde{F}$ be the data for coefficients (μ_a, D) and $(\tilde{\mu}_a, \tilde{D})$ respectively. Then there is an open set of illuminations $(q_1, q_2) \in (C^{1,\alpha}(\Gamma))^2$ for some $\alpha > \frac{1}{2}$ such that if $F = \tilde{F}$, then $(\mu_a, D) = (\tilde{\mu}_a, \tilde{D})$.*

Theorem 5.2. *Let $N_s = 2n$, $k \geq 2$ and Ω be an arbitrary bounded domain with C^{k+1} boundary Γ. Assume $(\mu_a, D), (\tilde{\mu}_a, \tilde{D}) \in V$ and $D = \tilde{D}$ on Γ. Let F and $\tilde{F}$ be the data for coefficients (μ_a, D) and $(\tilde{\mu}_a, \tilde{D})$ respectively. Then there is an open set of illuminations $(q_1, \cdots, q_{2n}) \in (C^{k,\alpha}(\Gamma))^{2n}$ and a constant C such that*

$$\|\mu - \tilde{\mu}_a\|_{C^k} + \|D - \tilde{D}\|_{C^k} \leq C\|F - \tilde{F}\|_{(C^{k+1})^{2n}}. \tag{5.17}$$

Both theorems can be proved by first using the Liouville transform $u = \sqrt{D}\phi$ to transform the DA (5.14) into the Schrödinger equation

$$\triangle u_i + \nu u_i = 0, \ \mathbf{x} \in \Omega,$$
$$u_i = \sqrt{D}q_i, \ \mathbf{x} \in \Gamma, \tag{5.18}$$

with

$$\nu = -\frac{\triangle\sqrt{D}}{\sqrt{D}} - \frac{\mu_a}{D}, \tag{5.19}$$

and then proving the equivalent theorems by complex geometric optics (CGO) solutions for the Schrödinger equation (5.18) with the internal data $\{\mu u_i, i \leq N_s\}$, where μ is defined as

$$\mu = \frac{\mu_a}{\sqrt{D}}. \tag{5.20}$$

Once it is shown for the Schrödinger (5.18) that (μ, ν) can be uniquely (resp. stably) reconstructed from $\{\mu u_i, i \leq N_s\}$, both theorems follow automatically for DA.

The assumption $N_s = 2n$ in Theorem 2 can be relaxed to $N_s = 2$ with additional geometric hypothesis on X. We refer the interested readers to [12] for this alternative stability estimate and the details of the proofs. Furthermore, the changes of variables (5.19) and (5.20) provide a constructive method for recovering (μ_a, D) as will further discussed next.

5.2.3 Algorithm

Given the data $Y = \{Y_{ij}, i \leq N_s, j \leq N_t\}$, the objective of QPAT is to obtain the optical maps X by minimizing the difference between the model prediction F and the actual data Y

$$f(X) = \frac{1}{2}||F(X) - Y||_2^2 = \frac{1}{2}\sum_{i=1}^{N_s}\sum_{j=1}^{N_t}(F_{ij}(X) - Y_{ij})^2. \tag{5.21}$$

For example, $X = \mu_a$ or $X = (\mu_a, \mu_s')$ in DA based QPAT; $X = \mu_a$ or $X = (\mu_a, \mu_s)$ in RTE based QPAT assuming g is known. Please note that the simultaneous reconstruction of both optical maps based on DA is not unique in the conventional setting with single optical illumination under single wavelength. Similarly, in the RTE based QPAT, since it seems very unlikely to extract singular components in the practical cases that are not fully in the transport regime, it should be more appropriate to consider RTE based MS-QPAT as well in order to reconstruct both optical maps. Therefore, in the following we consider the setting of MS-QPAT with $N_s > 1$.

Here the model prediction F is consistent with the previous definitions (5.6) or (5.13), i.e.,

$$F = \{F_{ij} := P_j^T \Phi_i, i \leq N_s, j \leq N_t\}, \tag{5.22}$$

where P_j is the detection functional defined by (5.6) for DA or (5.13) for RTE that consists of μ_a and interpolation weights, and Φ_i is the photon density solution from the ith illumination that can be obtained from solving linear systems $A\Phi_i = Q_i$ with the formulas (5.4) and (5.5) for DA model, and the formulas (5.11) and (5.12) for RTE model.

Fixed-Point Iteration

Assuming the knowledge of the scattering map, fixed-point iteration can be used to reconstruct μ_a, in which μ_a and Φ are sequentially updated in iterations. That is formally the following with the initial guess μ_a^0

$$\mu_a^{n+1} = \frac{Y}{\Phi(\mu_a^n)}, \tag{5.23}$$

where we implicitly assume $N_s = 1$, and the interpolation may be necessary for mapping Φ to the same discretized location of μ_a, i.e., with the interpolation weights in (5.6) or (5.13). In this simple fixed-point iteration scheme (5.23), the next solution μ_a^{n+1} is directly updated each time accompanied by solving a linear system for Φ with respect to the last solution μ_a^n.

Here the initial guess μ_a^0 has to be considerably close to the true distribution in order for μ_a to converge. Overall, it is sensitive to the initial guess μ_a^0, the insufficient knowledge of the scattering property, and the data noise.

DA Based Non-iterative Methods

With DA as forward model, QPAT can be solved in a non-iterative fashion. Before starting the discussion of non-iterative methods, it is important to notice that these methods can be non-iterative in the ideal setting, e.g., noise-free cases. In the practical setting, the non-iterative reconstruction can be problematic, for which the iterative regularization or optimization of the solution is usually required.

When the diffusion coefficient D or the reduced scattering coefficient μ_s' is assumed to be known, one can simplify the DA (5.1) by inserting the data $F = \mu_a \phi$ directly, i.e.,

$$-\nabla \cdot (D\nabla\phi) = -F, \ \in \Omega,$$
$$2AD(\mathbf{n} \cdot \nabla\phi) + \phi = q, \ \mathbf{x} \in \Gamma. \tag{5.24}$$

After solving (5.24) for ϕ, μ_a can be simply computed by

$$\mu_a = \frac{F}{\phi}. \tag{5.25}$$

In the second case that D is not known in Ω, (μ_a, D) can still be recovered non-iteratively [9] assuming at least two optical illuminations i.e., $\{F_i = \mu_a\phi_i, \ i = 1, 2\}$, the exact D in Γ and the following DA with Dirichlet boundary condition

$$-\nabla \cdot (D\nabla\phi_i) + \mu_a\phi_i = 0, \ \mathbf{x} \in \Omega,$$
$$\phi_i = q_i, \ \mathbf{x} \in \Gamma. \tag{5.26}$$

From Dirichlet boundary condition and the data F, μ_a is also exactly known in Γ, i.e., $\mu_a = F_i/q_i$. The algorithm goes as follow.

By multiplying DA (5.14) for ϕ_1 by ϕ_2, DA (5.26) for ϕ_2 by ϕ_1, subtracting two equations, we have

$$-\nabla \cdot (D\phi_1^2 \nabla\frac{\phi_2}{\phi_1}) = 0. \tag{5.27}$$

Using Liouville variable $u = \sqrt{D}\phi_1$, we obtain the first-order transport equation with the variable u^2

$$-\nabla \cdot (u^2 \nabla \tfrac{F_2}{F_1}) = 0, \ \mathbf{x} \in \Omega,$$
$$u^2 = Dq_1^2, \ \mathbf{x} \in \Gamma. \tag{5.28}$$

After solving (5.28) for u, we can obtain ν (5.19) and μ (5.20) by formulas

$$\mu = \frac{\mu_a}{\sqrt{D}} = \sqrt{\frac{(\mu_a\phi_1)^2}{D\phi_1^2}} = \frac{F_1}{u},$$
$$\nu = -\frac{\triangle\sqrt{D}}{\sqrt{D}} - \frac{\mu_a}{D} = -\frac{\triangle(\sqrt{D}\phi_1)}{\sqrt{D}\phi_1} = -\frac{\triangle u}{u}, \tag{5.29}$$

where DA (5.26) is used in deriving the second formula of (5.29).

Next observing that

$$\triangle\sqrt{D} + \nu\sqrt{D} = -\mu, \tag{5.30}$$

from (5.19) or the second formula of (5.29), we can compute $\sqrt{D}$ from ν and μ by solving the elliptic (5.30) with Dirichlet boundary condition, i.e., the exact values of $\sqrt{D}$ on Γ. Consequently, we can compute D, and $\mu_a = \mu\sqrt{D}$.

In summary, one can reconstruct (μ_a, D) from two measurements by solving a first-order transport (5.28) and an elliptic (5.30). Now we state one more assumption on F in order for the solution to make sense. That is

$$\beta := F_1^2 \nabla \frac{F_2}{F_1} \tag{5.31}$$

is a vector in $W^{1,\infty}$ and $|\beta| \geq \alpha_0 > 0$.

In implementation, since β usually varies significantly over the domain, the transport (5.28) is solved in the normalized vector field $|\beta|/\mu^2$ instead. In the practical case with noise, the derivative of F_2/F_1 can become problematic. One way to deal with the noise is to formulate (5.28) as a linear system and iteratively solve it using some proper regularization.

This non-iterative reconstruction method can be potentially appealing for its non-iterative nature. However, we will not adopt this approach here in designing our solver for 3D large-scale QPAT, mainly because (μ_a, D) is assumed to be known on Γ. Another reason is that it seems that the practical performance is data-dependent, such as β and its lower bound α_0.

Jacobian Based Method

In the Jacobian based method [26], the QPAT is formulated as a least-square problem (5.21) with the forward model F that is nonlinearly dependent on optical parameters X, and then the Jacobians of F are iteratively computed so that the consequent subproblems are linear least square problems that can be solved by various least-square or convex optimization techniques [13,24,33].

Jacobian based method is a natural way for solving (5.21) as an typical iterative linearization approach for nonlinear problems. That is the following with some initial guess X^0,

$$B^{n+1} = Y + J(X^n)X^n - F(X^n),$$
$$X^{n+1} = \text{argmin}_X ||J(X^n)X - B^{n+1}||_2^2 + R(X), \quad (5.32)$$

where $J(X^n)$ is the Jacobian computed at X^n, and $R(X)$ the regularization term to regularize the solution that can be based on some priors of X, such as smoothness and sparsity. Another consequence for the image regularization is that the minimization problems of (5.32) are usually less illposed. This can be interpreted as a result of finding solutions in a more restrictive space with the global minimum.

The main advantages of Jacobian based method (5.32) are its simplicity and flexibility. For example, many state-of-art image processing techniques can be potentially applied here. But, for large-scale QPAT in 3D, the computation of Jacobian J can be prohibitively slow, as will be explained next.

J can be computed as follow. First we differentiate both sides of (5.2) with respect to x_k, each component of X,

$$A\frac{\partial \Phi_i}{\partial x_k} = -\frac{\partial A}{\partial x_k}\Phi_i \quad (5.33)$$

and use it to compute the Jacobian in the following way

$$\begin{aligned}
[J]_{ij,k} &= \frac{\partial F_{ij}}{\partial x_k} \\
&= \frac{\partial P_j^T}{\partial x_k}\Phi_i + P_j^T\frac{\partial \Phi_i}{\partial x_k} \\
&= \frac{\partial P_j^T}{\partial x_k}\Phi_i - P_j^T[A^{-1}(\frac{\partial A}{\partial x_k}\Phi_i)] \\
&= \frac{\partial P_j^T}{\partial x_k}\Phi_i - (P_j^T A^{-1})(\frac{\partial A}{\partial x_k}\Phi_i) \\
&= \frac{\partial P_j^T}{\partial x_k}\Phi_i - (\Psi_j^T)(\frac{\partial A}{\partial x_k}\Phi_i),
\end{aligned} \quad (5.34)$$

where Ψ_j is so-called adjoint solution defined by

$$\Psi_j = (A^T)^{-1}P_j \text{ or } A^T\Psi_j = P_j. \quad (5.35)$$

As a result, this method for computing Jacobian is so-called the adjoint method, which is usually much more efficient than the direct method [1]. But, it still requires the computation of linear systems (5.2) for $N_s + N_t$ times, that can be extremely time-consuming for 3D large-scale QPAT since N_t can easily be around a million or more in 3D.

Gradient Based Method

An alternative way without computing Jacobians is to consider (5.21) as a fully nonlinear function instead of the least-square problem [26]. That is

$$X = \mathrm{argmin}_X f(X) + R(X), \tag{5.36}$$

where $R(X)$ again is the regularization on the solution X.

Now we only need to compute the gradient of $f(X)$. The adjoint method for gradient computation goes as follow

$$\begin{aligned}
[\partial f]_k &= \sum_{i,j} (F_{ij} - Y_{ij}) \frac{\partial F_{ij}}{\partial x_k} \\
&= \sum_{i,j} (F_{ij} - Y_{ij}) \frac{\partial P_j^T}{\partial x_k} \Phi_i + \sum_{i,j} (F_{ij} - Y_{ij}) P_j^T \frac{\partial \Phi_i}{\partial x_k} \\
&= \sum_{i,j} (F_{ij} - Y_{ij}) \frac{\partial P_j^T}{\partial x_k} \Phi_i - \sum_i S_i^T [A^{-1} (\frac{\partial A}{\partial x_k} \Phi_i)] \\
&= \sum_{i,j} (F_{ij} - Y_{ij}) \frac{\partial P_j^T}{\partial x_k} \Phi_i - \sum_i \Psi_i^T (\frac{\partial A}{\partial x_k} \Phi_i)
\end{aligned} \tag{5.37}$$

with the adjoint source

$$S_i = \sum_j (F_{ij} - Y_{ij}) P_j \tag{5.38}$$

and the adjoint solution

$$\Psi_i = (A^T)^{-1} S_i \text{ or } A^T \Psi_i = S_i. \tag{5.39}$$

Please note that the first term in (5.34) and (5.37) is the singular absorption term that is nonzero only for the absorption coefficient μ_a, which reflects the fact that the sensitivity with respect to μ_a is stronger than with respect to the other variable. Consequently the reconstruction of μ_a should have a better resolution. On the other hand, the aforementioned fixed-point iteration actually corresponds to the truncated Jacobian involving only the first term for the absorption coefficient. Therefore, fixed-point iteration usually fails when the first term is no longer dominating or with considerable errors.

Now, for each gradient we only need to compute linear systems (5.2) for $2N_s$ times that are only a few. In contrast with the method based on Jacobians (5.34), the method based on gradients (5.37) requires much fewer computations of linear systems, and thus is feasible for 3D QPAT. The gradient based method has been developed for QPAT by [14, 26]. It is also the base of GBSB algorithm for large-scale 3D QPAT that will be introduced next.

5.3 GBSB: An Algorithm for Large-Scale 3D QPAT

In this section, we will aim at solving large-scale 3D QPAT, e.g., to recover both absorption map and scattering map in the setting of MS-QPAT. In the following, QPAT will be formulated as a bound-constrained nonlinear minimization problem with the solution regularized by TV norm, and then the development of the solution algorithm is based on the Split Bregman method [18, 22, 34], namely, GBSB, gradient-based bound-constrained split Bregman method, which is an extension of our prior work [26].

5.3.1 Formulation

For model simplicity, we adopt the following setting that is commonly assumed for QPAT: with DA as the forward model, the objective is to reconstruct the absorption coefficient μ_a and the reduced scattering coefficient μ'_s from the absorbed energy $\mu_a\phi$, which is assumed to be known (e.g., through acoustic inversion).

In the setting of MS-QPAT with N_s optical illuminations, the available data Y are

$$Y = \{Y_{ij} := F_j(\mu_a, \phi_i) + \epsilon_{ij}, \ i \le N_s, j \le N_d\}, \tag{5.40}$$

where ϕ_i is the DA solution with the ith source, F the measuring functional defined by (5.6) that is linearly dependent on μ_a and ϕ_i, N_d the number of the detectors, and ϵ_{ij} the data noise.

The parameters to be recovered are

$$X = (\mu_a, \mu'_s). \tag{5.41}$$

Assuming that the data noise ϵ obeys the Gaussian distribution, we formulate the data fidelity of QPAT as the nonlinear least-square summation

$$f(X) = \frac{1}{2}||F(X) - Y||_2^2 = \frac{1}{2}\sum_{i,j}[F_j(\mu_a, \phi_i(\mu_a, \mu'_s)) - Y_{ij}]^2. \tag{5.42}$$

In this study, we specifically choose piecewise-constant discretization of X

$$X = \{(\mu_{a,i}, \mu'_{s,i}), i \le N_t\}, \tag{5.43}$$

and regularize it by TV norm

$$||MX||_1 = \sum_i |M_{i1}\mu_{a,i1} - M_{i2}\mu_{a,i2}| + |M_{i1}\mu'_{s,i1} - M_{i2}\mu'_{s,i2}|, i \le N_e. \tag{5.44}$$

Here N_e is the total number of element edges in 2D or element surfaces in 3D that are inside the domain (excluding the domain boundary). It can be shown by TV coarea formula that M_{i1} and M_{i2} correspond to the edge length or the surface area [25]. Please note that the triangulation is assumed to be conformal so that each internal edge or surface i is shared by exactly two elements $i1$ and $i2$.

Consequently QPAT can be formulated as the following bound-constrained nonlinear optimization problem

$$X = \text{argmin}_X f(X) + \lambda ||MX||_1, \text{ subject to } L \leq X \leq U. \tag{5.45}$$

Here λ is the regularizing parameter, and L (resp. U) is the lower (resp. upper) bound of X. Please note that (1) the use of TV regularization or simple bounds is for regularizing the illposed solution, particularly for μ'_s; (2) the enforced simple bounds are loose constraints for excluding some apparently undesired solutions (e.g., lower bound by zero and upper bound by a order of magnitude of the maximum), rather than tight a priori constraints targeting directly at the desired solutions.

5.3.2 Split Bregman Method

Since the simple bounds will be handled explicitly, now let us consider the unconstrained version of (5.45),

$$X = \text{argmin}_X f(X) + \lambda ||MX||_1. \tag{5.46}$$

The simple version of a typical approach for solving (5.46) is to iteratively solve

$$\begin{aligned} X^{n+1} &= \text{argmin}_X f(X) + \lambda_n ||MX||_1, \\ \lambda_{n+1} &= \mu \lambda_n, 0 < \mu < 1. \end{aligned} \tag{5.47}$$

That is, the iteration begins with the regularized solution, that is from a less ill-conditioned system due to the regularization, and converges to the true solution as the regularization diminishes. A major difficulty of (5.47) is that the system can be too ill-conditioned to be solvable for small regularization parameters after some iterations.

To resolve this difficulty, we begin with the Bregman distance of TV norm

$$D(X,Y) = \lambda ||MX||_1 - \lambda ||MY||_1 - < V(Y), X - Y >, \tag{5.48}$$

where V is a subgradient of the TV norm at Y.

Now instead of varying λ_n as in the continuation strategy (5.47), we fix the regularizing parameter λ, replace the TV norm $\lambda ||MX||_1$ by its Bregman distance $D(X,Y)$, and iteratively solve

$$X^{n+1} = \text{argmin}_X f(X) + D(X, X^n). \tag{5.49}$$

That is

$$\begin{aligned} X^{n+1} &= \text{argmin}_X f(X) + \lambda ||MX||_1 - V^n X, \\ V^{n+1} &= V^n - \partial f(X^{n+1}). \end{aligned} \tag{5.50}$$

(5.50) is the precedent of Split Bregman Method that was proposed in [34]. Please note that (5.50) avoids to solve ill-conditioned systems by fixing λ while updating the Bregman distance. We refer the readers to [18, 34] for the well-definedness, convergence and several nice properties of the Bregman method when $f(X)$ is convex, and [5] for the analysis of more general $f(X)$ commonly occurring in inverse problems.

The similar Bregman method can be used to solve the non-differentiable L_1-type subproblems of (5.50)

$$X = \mathrm{argmin}_X g^n(X) + \lambda ||MX||_1 \qquad (5.51)$$

with $g^n(X) = f(X) - V^n X$.

The motivation comes from (1) the fact that the L_1 scalar minimization

$$\min \frac{1}{2}(x - y)^2 + \lambda |x| \qquad (5.52)$$

has the explicit solution, so-called the shrinkage formula

$$x = T_\lambda(y) = \mathrm{sgn}(y) \cdot \max(|y| - \lambda, 0), \qquad (5.53)$$

where $\mathrm{sgn}(y)$ denotes the sign of the scalar y;(2) the split treatment of the differentiable data fidelity and the non-differentiable TV norm. Please note that (5.53) also extends to the vector computation since the objective function of (5.51) is separable into the summation of (5.52).

The method is so-called Split Bregman Method [18, 22] and it goes as follow. First we let the dummy variable $Z = MX$ so that TV becomes $||Z||_1$, for which the Shrinkage formula (5.53) can be applied later. Consequently, (5.51) is reformulated as an equality-constrained optimization problem

$$(X, Z) = \mathrm{argmin}_{(X,Z)} g^n(X) + \lambda ||Z||_1, \text{ subject to } MX = Z. \qquad (5.54)$$

Now we enforce the equality constraints by quadratic penalties, however penalize the Bregman distance of $g^n(X) + \lambda ||Z||_1$ iteratively rather than update the penalizing parameter μ, i.e.,

$$(X^{m+1}, Z^{m+1}) = \mathrm{argmin}_{(X,Z)} g^n(X) + \lambda ||Z||_1$$
$$- V_x^m X - V_z^m Z + \frac{1}{2}\mu ||MX - Z||_2^2,$$
$$V_x^{m+1} = V_x^m + \mu M^T(Z^{m+1} - MX^{m+1}),$$
$$V_z^{m+1} = V_z^m + \mu(MX^{m+1} - Z^{m+1}), \qquad (5.55)$$

which can be simplified to

$$(X^{m+1}, Z^{m+1}) = \mathrm{argmin}_{(X,Z)} g^n(X) + \lambda ||Z||_1 + \frac{1}{2}\mu ||MX - Z + W^m||_2^2,$$
$$W^{m+1}7 = W^m + MX^{m+1} - Z^{m+1}. \qquad (5.56)$$

Regarding the handling of equality constraints, Split Bregman Method (5.56) is similar to the augmented Lagrangian Method [4, 27, 33, 35], in which the Lagrangian multipliers are added to the object function and are iteratively estimated with the fixed quadratic penalty parameter μ.

Then the first step of (5.56) can be solved by the iterative alternating optimization of X and Z

$$
\begin{aligned}
X^{k+1} &= \mathrm{argmin}_X g^n(X) + \tfrac{1}{2}\mu\|MX - Z^k + W^m\|_2^2, \\
Z^{k+1} &= \mathrm{argmin}_Z \lambda\|Z\|_1 + \tfrac{1}{2}\mu\|MX^{k+1} - Z + W^m\|_2^2,
\end{aligned}
\tag{5.57}
$$

where the second equation has the explicit solution by shrinkage formula (5.53)

$$
Z^{k+1} = T_{\frac{\lambda}{\mu}}(MX^{k+1} + W^m).
\tag{5.58}
$$

Combining (5.50), (5.56), (5.57) and (5.58), Split Bregman Method is now summarized as

> For $n = 1$ to N
>> For $m = 1$ to M
>>> For $k = 1$ to K
>>>> $X^{k+1} = \mathrm{argmin}_X f(X) - V^n X + \tfrac{1}{2}\mu\|MX - Z^k + W^m\|_2^2;$
>>>> $Z^{k+1} = T_{\frac{\lambda}{\mu}}(MX^{k+1} + W^m);$
>>> End
>>> $W^{m+1} = W^m + MX^{m+1} - Z^{m+1};$
>> End
>> $V^{n+1} = V^n - \partial f(X^{n+1}).$
> End

In particular, this Split Bregman loop with $M = K = 1$ has been proven to converge with certain numerical advantages [18] and will be adopted as the base of the GBSB algorithm. That is

$$
\begin{aligned}
X^{n+1} &= \mathrm{argmin}_X f(X) + \frac{1}{2}\mu\|MX - Z^n + W^n\|_2^2 - V^n X, \\
Z^{n+1} &= T_{\frac{\lambda}{\mu}}(MX^{n+1} + W^n), \\
W^{n+1} &= W^n + MX^{n+1} - Z^{n+1}, \\
V^{n+1} &= V^n - \partial f(X^{n+1}).
\end{aligned}
\tag{5.59}
$$

All steps in (5.59) except the first one are computationally cheap. Next we will turn to the discussion for solving this differentiable subproblem to update X.

5.3.3 L-BFGS: Quasi-Newton Approximation of the Hessian

Since the computation of Jacobians is extremely time-consuming, we will not formulate Jacobians from (5.21) according to the least-square form of $f(X)$, and solve the consequent optimization problems using least-square techniques. Instead, we rather treat (5.21) as a nonlinear objective function, and only compute its gradients (please see Sect. 5.2.3 for the details of gradient computation).

Specifically, we will adopt the Quasi-Newton method with adaptive updating of the Hessian by gradients (Sect. 5.3.3), compute the search direction while explicitly enforcing simple bounds (Sect. 5.3.4), and perform the line search satisfying Wolfe conditions and simple bounds (Sect. 5.3.5).

We are going to focus on solving the first subproblem of the Split Bregman loop (5.59). To simplify the notation, we let

$$g(X) = f(X) + \frac{1}{2}\mu\|MX - Z^n + W^n\|_2^2 - V^n X \qquad (5.60)$$

and consider

$$X = \operatorname{argmin}_X g(X). \qquad (5.61)$$

Suppose that the current iterate is X_k, a quadratic approximation of (5.61) with $p = X - X_k$ is

$$Q_k(p) = g_k + \partial g_k^T p + \frac{1}{2}p^T H_k p, \qquad (5.62)$$

where g_k, ∂g_k and H_k are the function value, the gradient and the Hessian at X_k respectively. Consequently, from the optimal condition of (5.62),

$$p_k = -H_k^{-1}\partial g_k, \qquad (5.63)$$

which can be used as the search direction at the kth iteration for (5.61).

For computational efficiency, instead of formulating H_k explicitly and taking its inverse H_k^{-1}, we shall use a well-known Quasi-Newton method, namely, BFGS method [20, 21, 33], to iteratively update H_k^{-1}, for which only ∂g_k is required. The methodology goes as follow.

Let p_k and α_k be the search direction and the step length at the current iterate X_k, i.e.,

$$X_{k+1} = X_k + \alpha_k p_k. \qquad (5.64)$$

At the next iterate X_{k+1}, the new quadratic approximation $Q_{k+1}(p)$ should be consistent in the sense that

$$\begin{aligned}\partial Q_{k+1}(-\alpha_k p_k) &= \partial g_k, \\ \partial g_{k+1} + H_{k+1}(-\alpha_k p_k) &= \partial g_k.\end{aligned} \qquad (5.65)$$

That is

$$s_k = H_{k+1}^{-1} y_k \qquad (5.66)$$

with

$$s_k = x_{k+1} - x_k \text{ and } y_k = \partial g_{k+1} - \partial g_k. \qquad (5.67)$$

To uniquely update H_{k+1}^{-1} from H_k^{-1}, besides the symmetric requirement and the secant condition (5.66), we impose the additional condition that H_{k+1}^{-1} is the closest to H_k^{-1} in the weighted Frobenius norm $|| \cdot ||$ [33], i.e.,

$$H_{k+1}^{-1} = \mathrm{argmin}_{H^{-1}} ||H^{-1} - H_k^{-1}|| \\ \text{subject to } H^{-1} = H^{-1^T} \text{ and } H^{-1}y_k = s_k. \tag{5.68}$$

The unique solution of (5.68) is the well-known BFGS formula

$$H_{k+1}^{-1} = V_k^T H_k^{-1} V_k + \rho_k s_k s_k^T \tag{5.69}$$

with

$$\rho_k = \frac{1}{y_k^T s_k} \text{ and } V_k = I - \rho_k y_k s_k^T. \tag{5.70}$$

However, BFGS formula (5.69) is still not suitable for large-scale computation since H_k^{-1} is usually dense and consequently can be prohibitive in terms of memory and speed. Therefore, we adopt the limited-memory version of BFGS, so-called L-BFGS [8,33].

The motivation of L-BFGS comes from a recursive reformulation of (5.69)

$$H_k^{-1} = (V_{k-1}^T \dots V_0^T) H_0^{-1}(V_0 \dots V_{k-1}) + \rho_1(V_{k-1}^T \dots V_1^T) s_1 s_1^T (V_1 \dots V_{k-1}) \\ + \dots + \rho_{k-1} s_{k-1} s_{k-1}^T. \tag{5.71}$$

Now, for L-BFGS, we only save and use the most recent m pairs of (s,y) for updating H_k^{-1}, i.e.,

$$H_k^{-1} = (V_{k-1}^T \dots V_{k-m}^T) H_{k,0}^{-1}(V_{k-m} \dots V_{k-1}) \\ + \rho_{k-m}(V_{k-1}^T \dots V_{k-m+1}^T) s_{k-m} s_{k-m}^T (V_{k-m+1} \dots V_{k-1}) \\ + \dots + \rho_{k-1} s_{k-1} s_{k-1}^T, \tag{5.72}$$

where $H_{k,0}^{-1}$ is the initial guess of H^{-1} at the kth iteration. An empirical effective choice is

$$H_{k,0}^{-1} = \gamma_k I \text{ with } \gamma_k = \frac{s_{k-1}^T y_{k-1}}{y_{k-1}^T y_{k-1}}. \tag{5.73}$$

This L-BFGS recursive formula (5.72) allows an efficient computation of the search direction p_k at the current iterate X_k, i.e.,

L-BFGS Update: $p_k = \mathrm{LBFGS}(\partial g_k, \{s_i, y_i, k - m \le i \le k - 1\})$.
    $q = \partial g_k$;
    For $i = k - 1$ to $k - m$
        $q = q - \alpha_i y_i$;
    End
    $r = H_{k,0}^{-1} q$;

For $i = k - m$ to $k - 1$
$$r = r + s_i(\alpha_i - \rho_i y_i^T r);$$
End
$$p_k = -r.$$

We find that the L-BFGS (5.72) with around five truncated terms ($m = 5$) is generally sufficient in Q-PAT. We refer the readers to [20, 21, 33] and the references therein for the general framework of Quasi-Newton methods, convergence analysis of BFGS (5.69) and L-BFGS (5.72), and other useful updates of Hessian.

5.3.4 Bound-Constrained Search Direction

Now let us come back to the original constrained QPAT problem (5.45) and explicitly deal with its simple bounds. In this case, the Split Bregman loop (5.59) still works and instead we consider its first step as

$$X = \operatorname{argmin}_X g(X) \text{ subject to } L \leq X \leq U \tag{5.74}$$

with $g(X)$ defined by (5.60).

In GBSB, we explicitly enforce bound constraints in the computation of search directions of (5.74) from the first-order necessary conditions (KKT conditions) [33]. The similar explicit formula for simple bounds was also implemented in the nonlinear optimization software LANCELOT [16]. That is, assuming $p(X)$ is the exact decent direction of (5.61), it is not difficult to verify that the projected gradient $\hat{p}(X)$ onto the feasible region is exactly

Projection of p onto the Feasible Region: $\hat{p}_k = \operatorname{Proj}(p_k)$.

$$\hat{p}(x) = \begin{cases} -L(x), & \text{if } x + p(x) \leq L(x) \\ p(x), & \text{if } L(x) < x + p(x) < U(x), \ x \in X \\ -U(x), & \text{if } x + p(x) \geq U(x) \end{cases} \tag{5.75}$$

(5.75) is the updating formula of search directions in the Quasi-Newton method. That is, the search direction $p(X)$ is first computed by L-BFGS (5.72), and then projected to $\hat{p}(X)$ via (5.75). It is $\hat{p}(X)$ that will be used in line search for the constrained QPAT (5.45). Please note that $\hat{p}(X)$ is not exact in our Quasi-Newton method since $p(X)$ is from L-BFGS in iterative quadratic approximations of f. Therefore, we will also enforce the bound constraints in the line search.

5.3.5 Line Search

Now we are going to complete the solution method for minimizing the nonlinear functional (5.74) as subproblems during Split Bregman iterations (5.59) for the QPAT model problem (5.45). The discussion concerns the choice of the step length α_k along the current search direction $\hat{p}_k$ so that new iterate X_{k+1} can be updated by

$$X_{k+1} = X_k + \alpha_k \hat{p}_k. \tag{5.76}$$

In GBSB, we enforce the Wolfe Conditions [33] for selecting α_k, i.e.,

$$\begin{aligned} g(X_k + \alpha_k \hat{p}_k) &\leq g(X) + c_1 \alpha_k \partial g^T(X_k)\hat{p}_k \\ \partial g^T(X_k + \alpha_k \hat{p}_k)\hat{p}_k &\geq c_2 \partial g^T(X_k)\hat{p}_k, \end{aligned} \tag{5.77}$$

where the first inequality guarantees the sufficient decrease of the objective function value, and the second inequality, so-called curvature condition, rules out unacceptably short steps. Please note that the curvature condition is necessary to guarantee the positive curvature, (i.e., $s_k^T y_k > 0$), which is implied by the secant equation (5.66). Otherwise, the performance of L-BFGS method may degrade due to the violation of the curvature condition.

Besides, we also enforce the bound constraints, i.e.,

$$L \leq X_k + \alpha_k \hat{p}_k \leq U. \tag{5.78}$$

As mentioned earlier, our projected gradient $\hat{p}$ is only an approximation, and it may not fulfill the simple bounds. As a result, (5.78) is necessary for X_{k+1} to stay feasible.

Last, we implement the following Backtracking line search algorithm in GBSB, which is simple, yet empirically sufficient.

Backtracking line search: $\alpha_k = \text{Backtrack}(\hat{p}_k)$.
  Choose $c_1 = 0.0001, c_2 = 0.9, \rho = 0.5, \alpha_k = 1$;
  Do $\alpha_k = \rho \alpha_k$
  Until α_k satisfies (5.77) and (5.78).

One observation from using this backtracking ling search algorithm is that $\alpha_k = 1$ is usually accepted after one or a few Quasi-Newton iterations (5.76).

5.3.6 GBSB Algorithm

Now, we can summarize our discussions so far on the GBSB algorithm for QPAT (5.45) as follow.

For $n = 1$ to N (Bregman outer loop)

  $g^n = f(X) - V^n X + \frac{1}{2}\mu||MX - Z^n + W^n||_2^2$;
  $X_0^n = X^{n-1}$;
  For $k = 1$ to K (Quasi-Newton inner loop)
    $p_k = \text{LBFGS}(\partial g_k^n, \{s_i^n, y_i^n, k - m \leq i \leq k - 1\})$;

$$\hat{p}_k = \text{Proj}(p_k);$$
$$\alpha_k = \text{Backtrack}(\hat{p}_k);$$
$$X_{k+1}^n = X_k^n + \alpha_k \hat{p}_k;$$
$$\text{Break, if } ||g_k^n|| \leq \epsilon;$$

$\quad$ End
$$X^n = X_K^n;$$

$$Z^{n+1} = T_{\frac{\lambda}{\mu}}(MX^{n+1} + W^n);$$
$$W^{n+1} = W^n + MX^{n+1} - Z^{n+1};$$
$$V^{n+1} = V^n - \partial f(X^{n+1}).$$

End

As discussed earlier, we fix the value of the regularizing parameter λ to be a constant. Regarding the choice of this constant, any value of λ other than a particularly small number is sufficient although the considerably large value of λ may require extra iteration steps. Besides, $\mu = \lambda$ is recommended from our numerical experiences.

The stopping criterion for the Quasi-Newton inner loop of the GBSB algorithm is based on $\epsilon = \epsilon_i g_0^n$ with a small constant ϵ_0, e.g., $\epsilon_i = 0.0001$. This g_0^n-dependent stopping criterion is motivated by the scale variation of g^n in the the inner loop. On the other hand, the stopping criterion of Bregman outer loop is based on the difference between two consecutive iterative solutions, i.e., $||X^{n+1} - X^n|| < \epsilon_0 ||X^n||$,e.g., $\epsilon_i = 0.01$. Please note that in the case with considerably noisy data (5.40), the small ϵ_i is not recommended since X^n may otherwise converge to the solution corresponding to the noisy data, which is not desirable.

We refer the readers to the previous sections for the details of individual steps of the GBSB algorithm, such as the algorithm parameters. Next we are going to discuss on some practical factors that can be of great importance for the performance of GBSB algorithm in QPAT.

5.3.7 Data Scaling

The first scaling is with respect to the data fidelity term (5.21), which has considerably scale variation among different spatial elements. To balance the minimization of the inhomogeneous discrepancy between the model and the data, we can either consider the weighted data fidelity term with weights Y_{ij}^{-1}

$$f_1(X) = \frac{1}{2} \sum_{i,j} [Y_{ij}^{-1} F_j(\mu_a, \phi_i(\mu_a, \mu_s')) - 1]^2, \tag{5.79}$$

or the data fidelity term in logarithms

$$f_2(X) = \frac{1}{2} \sum_{i,j} [\log F_j(\mu_a, \phi_i(\mu_a, \mu_s')) - \log Y_{ij}]^2. \tag{5.80}$$

5.3.8 Parameter Scaling

The second scaling is with respect to the reconstruction variables (5.41), since μ_a and μ'_s usually differ from each other by 1 or 2 orders of magnitude. That is, now we consider the minimization problem with respect to the scaled parameters

$$X' = (\mu_a, r\mu'_s),\tag{5.81}$$

with r as a scaling constant that can be set to the ratio of the mean of the initial absorption coefficient μ_a^0 over the mean of the initial reduced scattering coefficient μ'^0_s.

Consequently, the gradient of f with respect to X' is

$$\partial_{X'} f = (\partial_{\mu_a} f, r^{-1}\partial_{\mu'_s} f).\tag{5.82}$$

5.3.9 Initial Guess

Due to the nonlinear nature of QPAT and its ill-posedness besides the fact that the problem may have many local minimizers, in order to reconstruct the desired optical parameters, almost exclusively, any iterative algorithm requires an educational initial guess X^0. The extensive numerical tests on the stability of GBSB with respect to X^0 indicate that GBSB is quite stable even if the difference between X^0 and the underground truth is considerably large, which is ideal for the large-scale QPAT. Nonetheless, a reasonable X^0 is still required for GBSB to converge to the meaningful solution. An empirical strategy for X^0 is the model based fitting of the homogeneous optical background from the data, e.g.,

$$X^0 = \mathrm{argmin}_X f(X).\tag{5.83}$$

Please note that here $X^0 = (\mu_a^0, \mu'^0_s)$ and is spatially independent so that (5.83) can be solved much more efficiently in comparison with the QPAT model problem (5.45).

5.3.10 Simulation Settings

We performed the GBSB algorithm for MS-QPAT with DA as forward model in both 2D and 3D.

In the 2D simulation, the phantom was based on the embedding of a 2D Shepp-Logan phantom into a circular domain of a 50 mm diameter and the center (0,0). In addition, the absorption coefficient map (Fig. 5.1a) had another four 1.25 mm-diameter circular inclusions centered at (−5,0), (−10,0), (−15,0) and (−20,0); the reduced scattering coefficient map (Fig. 5.1a) had another four 1.25 mm-diameter circular inclusions centered at (5,0), (10,0), (15,0) and (20,0). To mimic the MS-QPAT, the phantom was illuminated respectively with the optical sources at the boundary located in each quadrant

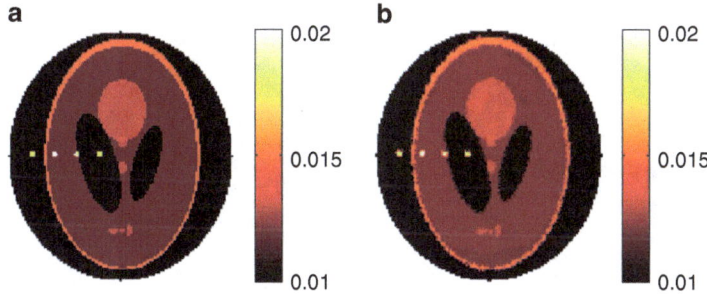

Fig. 5.1. 2D simultaneous reconstruction of absorption coefficient μ_a and reduced scattering coefficient μ'_s by the GBSB algorithm for MS-QPAT. (**a**) the true μ_a and (**b**) the reconstructed μ_a. The displays are 128×128 with the display window [0.01 0.02]. The reconstruction data has 1% Gaussian noise

of the circular domain, and the total number of optical illuminations was four in 2D.

To alleviate the inverse crime in simulations, the data Y (5.40) was generated based on a mesh with 17489 nodes and 34592 elements, and the reconstruction was performed on another independent mesh with 17137 nodes and 33888 elements. Then 1% Gaussian noise that is proportional to Y was added to the data before the reconstruction. Please note that the phantom was first generated on a 128×128 grid, and then interpolated to the 1st triangular mesh for generating Y. Therefore, the reconstruction phantom was the latter one after the interpolation. In this case, the resolution of the reconstruction phantom can be regarded as 128×128 since the number of variables in the piecewise-constant discretization was sufficiently large. On the other hand, the optical parameters were first reconstructed on the 2nd triangular mesh, and then interpolated to the 128×128 cartesian grid for display, which corresponded to the $< 0.5\,\mathrm{mm}$ resolution in the reconstruction phantom. On the other hand, as discussed earlier, the GBSB algorithm is stable with respect to the initial guess. The displayed reconstructed maps (Fig. 5.1b and 5.2b) were from a typical initial guess with $\mu_a^0 = 0.008$ and $\mu'^0_s = 0.8$, which did not overlap with any pixel value in the phantom (Figs. 5.1a and 5.2a).

Similarly, in the 3D simulation, the phantom was based on the embedding of a 3D Shepp-Logan phantom into a cylindrical domain of a 50 mm diameter and a 50 mm height and the center (0,0,0). In addition, the absorption coefficient map (Fig. 5.3a–c) had another four 2.5 mm-diameter spherical inclusions centered at (−5,0,0), (−10,0,0), (−15,0,0) and (−20,0,0); the reduced scattering coefficient map (Fig. 5.4a–c) had another four 2.5 mm-diameter spherical inclusions centered at (5,0,0), (10,0,0), (15,0,0) and (20,0,0). To mimic the MS-QPAT, the phantom was illuminated with the optical sources respectively from the top flat surface, the bottom flat surface, and then at the boundary located in each quadrant of the side, and the total number of optical illuminations was six in 3D.

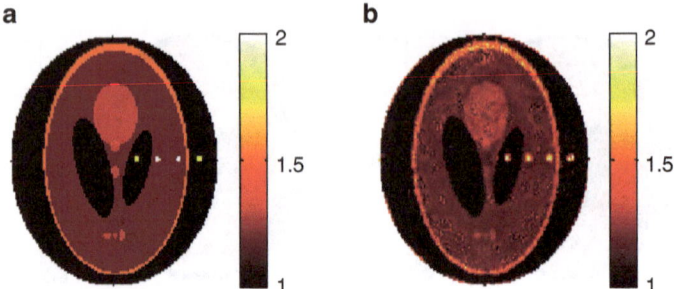

Fig. 5.2. 2D simultaneous reconstruction of absorption coefficient μ_a and reduced scattering coefficient μ_s' by the GBSB algorithm for MS-QPAT. (**a**) the true μ_s' and (**b**) the reconstructed μ_s'. The displays are 128×128 with the display window [1 2]. The reconstruction data has 1% Gaussian noise

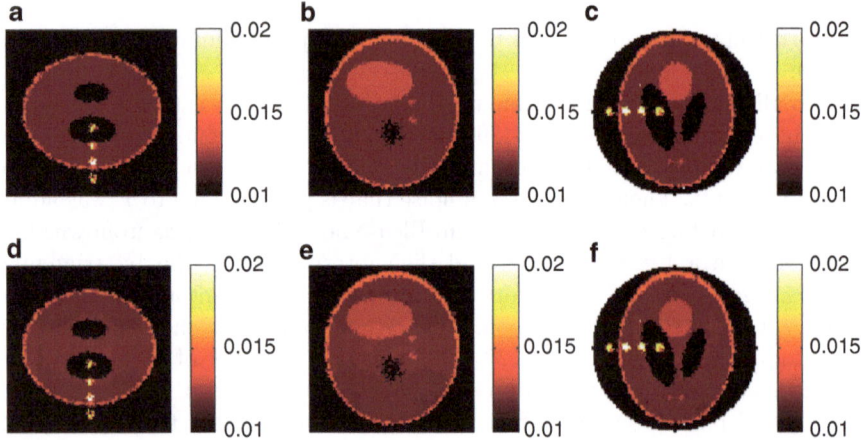

Fig. 5.3. 3D simultaneous reconstruction of absorption coefficient μ_a and reduced scattering coefficient μ_s' by the GBSB algorithm for MS-QPAT. (**a–c**) the sagittal, coronal, and transverse plane of the true μ_a and (**d–f**) the sagittal, coronal, and transverse plane of the reconstructed μ_a. The displays are $100 \times 100 \times 100$ with the display window [0.01 0.02]. The reconstruction data has 1% Gaussian noise

To alleviate the inverse crime in simulations, the data Y was generated based on a mesh with 64356 nodes and 446823 elements, and the reconstruction was performed on another independent mesh with 56635 nodes and 348687 elements. Then 1% Gaussian noise that is proportional to Y was added to the data before the reconstruction. Again the phantom was first generated on a $100 \times 100 \times 100$ grid, and then interpolated to the 1st tetrahedral mesh for generating Y. Please note that the resolution of the reconstruction phantom was actually less than $100 \times 100 \times 100$ due to the 1st tetrahedral mesh, i.e., $446823 < 100^3$. On the other hand, the optical parameters were first reconstructed on the 2nd tetrahedral mesh, and then interpolated to the $100 \times 100 \times 100$ cartesian grid for display, which corresponded to 0.5 mm resolution in the reconstruction phantom. Again, the actual reconstructed resolution did not reach $100 \times 100 \times 100$ since the number of variables in the 2nd tetrahedral mesh was smaller than 100^3 besides the fact that the reconstruction phantom did not even have the $100 \times 100 \times 100$ resolution. It was estimated that the reconstruction solution was approximately $70 \times 70 \times 70$. On the other hand, as discussed earlier, the GBSB algorithm is stable with respect to the initial guess. The displayed reconstructed maps (Fig. 5.3d–f and 5.4d–f) were from a typical initial guess with $\mu_a^0 = 0.008$ and $\mu'_s{}^0 = 0.8$, which did not overlap with any pixel value in the phantom (Figs. 5.3a–c and 5.4a–c).

5.3.11 Simulation Results

In this section, we present the simultaneously reconstructed absorption coefficient map μ_a and reduced scattering coefficient map μ'_s from the GBSB algorithm for MS-QPAT with DA as forward model.

The 2D results are displayed with the 128×128 resolution in Fig. 5.1 and 5.2: Fig. 5.1 shows the true and the reconstructed μ_a; Fig. 5.2 shows the true and the reconstructed μ'_s. The phantom is 50 mm in diameter, and therefore the image has < 0.5 mm resolution. In Fig. 5.1b, all Shepp-Logan inclusions and all four 1.25 mm inclusions of the μ_a phantom (Fig. 5.1a) are successfully reconstructed. In Fig. 5.2b, major Shepp-Logan inclusions and all four 1.25 mm inclusions of the μ'_s phantom (Fig. 5.1a) are successfully reconstructed while the small Shepp-Logan features are blurred and the background is relatively noisy. This reconstruction difference is due to the fact that the data are more sensitive to μ_a or the reconstruction is more ill-posed in μ'_s.

The 3D results are displayed with the $100 \times 100 \times 100$ resolution in Fig. 5.3 and 5.4: Fig. 5.3 shows the true and the reconstructed μ_a; Fig. 5.4 shows the true and the reconstructed μ'_s. The phantom is 50 mm in diameter and height, and therefore the image has < 0.5 mm resolution. However, as discussed earlier, the actual resolution is around $70 \times 70 \times 70$ due to the resolution of the used mesh. A better resolution should be achieved with finer meshes. In Fig. 5.3d–f, all Shepp-Logan inclusions and all four 2.5 mm inclusions of the μ_a phantom (Fig. 5.3a–c) are successfully reconstructed. Again, in Fig. 5.4d–f, major Shepp-Logan inclusions and all four 2.5 mm inclusions of the μ'_s phantom (Fig. 5.4a–c) are successfully reconstructed while the small Shepp-Logan features are blurred and the background is relatively noisy.

5.4 Discussion

The proposed GBSB algorithm can be easily extended for multi-wavelength QPAT [14] and RTE based QPAT. In the latter case, it is interesting to investigate whether it is possible to reconstruct (μ_a, μ_s, g) in the hybrid regime of

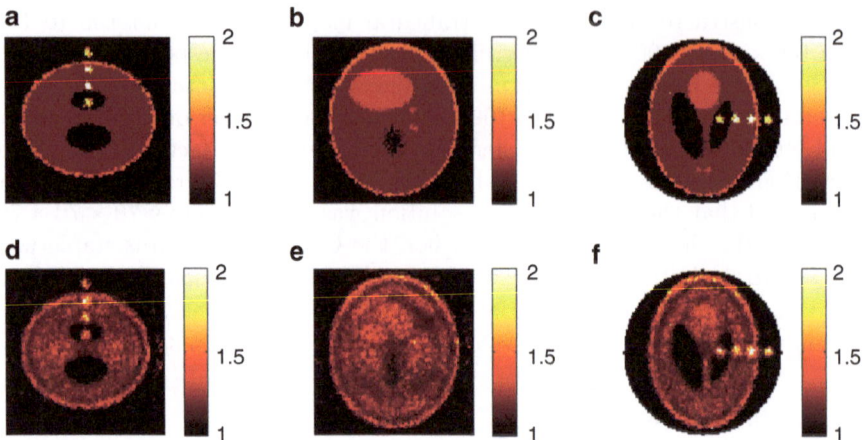

Fig. 5.4. 3D simultaneous reconstruction of absorption coefficient μ_a and reduced scattering coefficient μ'_s by the GBSB algorithm for MS-QPAT. (**a–c**) the sagittal, coronal, and transverse plane of the true μ'_s and (**d–f**) the sagittal, coronal, and transverse plane of the reconstructed μ'_s. The displays are $100 \times 100 \times 100$ with the display window [1 2]. The reconstruction data has 1% Gaussian noise

transport and diffusion. Another interesting question is to study the direct reconstruction of optical maps from boundary acoustic measurement, since GBSB can also be extended to this case and the forward/adjoint problem can be defined similarly once a practical acoustic model is available.

References

1. S.R. Arridge, Optical tomography in medical imaging. Inverse Probl. **15**, R41–R93 (1999)
2. M.L. Adams, E.W. Larsen, Fast iterative methods for discrete-ordinates particle transport calculations. Progr. Nucl. Energ. **40**, 3–159 (2002)
3. H. Ammari, E. Bossy, V. Jugnon, H. Kang, Reconstruction of the optical absorption coefficient of a small absorber from the absorbed energy density. SIAM J. Appl. Math. **71**, 676–693 (2011)
4. D.R. Bertsekas, Constrained Optimization and Lagrange Multiplier Methods. Academic Press, New York (1982)
5. M. Bachmayr, M. Burger, Iterative total variation schemes for nonlinear inverse problems. Inverse Probl. **25**, 105004 (2009)
6. B. Banerjee, S. Bagchi, R.M. Vasu, D. Roy, Quantitative photoacoustic tomography from boundary pressure measurements: noniterative recovery of optical absorption coefficient from the reconstructed absorbed energy map. J. Opt. Soc. Am. A **25**, 2347–2356 (2008)
7. G. Bal, A. Jollivet, V. Jugnon, Inverse transport theory of photoacoustics. Inverse Probl. **26**, 025011 (2010)

8. R.H. Byrd, J. Nocedal, R.B. Schnabel, Representations of quasi-Newton matrices and their use in limited-memory methods. Math. Program. Series A **63**, 129–156 (1994)

9. G. Bal, K. Ren, Multiple-source quantitative photoacoustic tomography in a diffuse regime. Inverse Probl. **27**, 075003 (2011)

10. Bal, G., Ren, K.: Non-uniqueness result for a hybrid inverse problem. Preprint (2010)

11. G. Bal, J.C. Schotland, Inverse scattering and acousto-optic imaging. Phys. Rev. Lett. **104**, 043902 (2010)

12. G. Bal, G. Uhlmann, Inverse diffusion theory of photoacoustics. Inverse Probl. **26**, 085010 (2010)

13. S. Boyd, L. Vandenberghe, Convex Optimization. Cambridge University Press, New York (2004)

14. B.T. Cox, S.R. Arridge, P.C. Beard, Estimating chromophore distributions from multiwavelength photoacoustic images. J. Opt. Soc. Am. A **26**, 443–455 (2009)

15. B.T. Cox, S.R. Arridge, K. Köstli, P. Beard, Quantitative photoacoustic imaging: fitting a model of light transport to the initial pressure distribution. Proc. SPIE **5697**, 49–55 (2005)

16. A.R. Conn, N.I.M. Gould, P.L. Toint, LANCELOT: A FORTRAN Package for Large-scale Nonlinear Optimization (Release A). Springer, New York (1992)

17. B. Cockburn, G.E. Karniadakis, C.W. Shu, Discontinuous Galerkin Methods: Theory, Computation and Applications. Springer, New York (2000)

18. J.-F. Cai, S. Osher, Z. Shen, Split Bregman methods and frame based image restoration. SIAM Multiscale Model. Simul. **8**, 337–369 (2009)

19. K.M. Case, P.F. Zweifel, Linear Transport Theory. Addison-Wesley, Massachusetts (1967)

20. J.E. Dennis, R.B. Schnabel, Numerical Methods for Unconstrained Optimization and Nolinear Equations. Prentice-Hall, Englewood Cliffs, NJ (1983)

21. R. Fletcher, Practical Methods of Optimization. Wiley, New York (1987)

22. T. Goldstein, S. Osher, The split Bregman method for l_1 regularized problems. SIAM J. Imaging Sci. **2**, 323–343 (2009)

23. H. Gao, H.K. Zhao, A fast forward solver of radiative transfer equation. Transport Theor. Stat. Phys. **38**, 149–192 (2009)

24. H. Gao, H.K. Zhao, Analysis of a forward solver of radiative transfer equation. Preprint (2010)

25. H. Gao, H. Zhao, Multilevel bioluminescence tomography based on radiative transfer equation Part 2: total variation and l1 data fidelity. Optics Express **18**, 2894–2912 (2010)

26. H. Gao, H.K. Zhao, S. Osher, Bregman methods in quantitative photoacoustic tomography. CAM Report **10–42**, (2010)

27. M.R. Hestenes, Multiplier and gradient methods. J. Optim. Theor. Appl. **4**, 303–320 (1969)

28. C. Johnson, Numerical Solution of Partial Differential Equations by the Finite Element Method. Cambridge University Press, New York (1987)

29. J.D. Jackson, Classical Electrodynamics. Wiley, New York (1999)

30. E.E. Lewis, W.F. Miller, Computational Methods of Neutron Transport. ANS Inc., La Grange Park, Illinois (1993)

31. C.H. Li, L.V. Wang, Photoacoustic tomography and sensing in biomedicine. Phys. Med. Biol. **54**, R59–R97 (2009)

32. NVIDIA: NVIDIA CUDA Compute Unified Device Architecture, Programming Guide version 2.2. (2009)
33. J. Nocedal, S.J. Wright, Numerical Optimization. Springer, New York (2006)
34. S. Osher, M. Burger, D. Goldfarb, J. Xu, W. Yin, An iterated regularization method for total variation based image restoration. SIAM Multiscale Model. Simul. **4**, 460–489 (2005)
35. M.J.D. Powell, A method for nonlinear constraints in minimization problems. In: R. Fletcher, (ed) Optimization. Academic Press, New York (1969)
36. J. Ripoll, V. Ntziachristos, Quantitative point source photoacoustic inversion formulas for scattering and absorbing media. Phys. Rev. E **71**, 031912 (2005)
37. L. Rudin, S. Osher, E. Fatemi, Nonlinear total variation based noise removal algorithms. J. Phys. D **60**, 259–268 (1992)
38. L.V. Wang, Multiscale photoacoustic microscopy and computed tomography. Nat. Photonics **3**, 503–509 (2009)
39. M. Xu, L.V. Wang, Photoacoustic imaging in biomedicine. Rev. Sci. Instrum. **77**, 041101 (2006)
40. L. Yin, Q. Wang, Q. Zhang, H. Jiang, Tomographic imaging of absolute optical absorption coefficient in turbid media using combined photoacoustic and diffusing light measurements. Opt. Lett. **32**, 2556–2558 (2007)
41. R.J. Zemp, Quantitative photoacoustic tomography with multiple optical sources. Appl. Optic. **49**, 3566–3572 (2010)

Index

H. Ammari (ed.), *Mathematical Modeling in Biomedical Imaging II*,
Lecture Notes in Mathematics 2035, DOI 10.1007/978-3-642-22990-9,
© Springer-Verlag Berlin Heidelberg 2012

LECTURE NOTES IN MATHEMATICS

Edited by J.-M. Morel, B. Teissier; P.K. Maini

Editorial Policy (for Multi-Author Publications: Summer Schools / Intensive Courses)

1. Lecture Notes aim to report new developments in all areas of mathematics and their applications - quickly, informally and at a high level. Mathematical texts analysing new developments in modelling and numerical simulation are welcome. Manuscripts should be reasonably selfcontained and rounded off. Thus they may, and often will, present not only results of the author but also related work by other people. They should provide sufficient motivation, examples and applications. There should also be an introduction making the text comprehensible to a wider audience. This clearly distinguishes Lecture Notes from journal articles or technical reports which normally are very concise. Articles intended for a journal but too long to be accepted by most journals, usually do not have this "lecture notes" character.

2. In general SUMMER SCHOOLS and other similar INTENSIVE COURSES are held to present mathematical topics that are close to the frontiers of recent research to an audience at the beginning or intermediate graduate level, who may want to continue with this area of work, for a thesis or later. This makes demands on the didactic aspects of the presentation. Because the subjects of such schools are advanced, there often exists no textbook, and so ideally, the publication resulting from such a school could be a first approximation to such a textbook. Usually several authors are involved in the writing, so it is not always simple to obtain a unified approach to the presentation.

    For prospective publication in LNM, the resulting manuscript should not be just a collection of course notes, each of which has been developed by an individual author with little or no coordination with the others, and with little or no common concept. The subject matter should dictate the structure of the book, and the authorship of each part or chapter should take secondary importance. Of course the choice of authors is crucial to the quality of the material at the school and in the book, and the intention here is not to belittle their impact, but simply to say that the book should be planned to be written by these authors jointly, and not just assembled as a result of what these authors happen to submit.

    This represents considerable preparatory work (as it is imperative to ensure that the authors know these criteria before they invest work on a manuscript), and also considerable editing work afterwards, to get the book into final shape. Still it is the form that holds the most promise of a successful book that will be used by its intended audience, rather than yet another volume of proceedings for the library shelf.

3. Manuscripts should be submitted either online at www.editorialmanager.com/lnm/ to Springer's mathematics editorial, or to one of the series editors. Volume editors are expected to arrange for the refereeing, to the usual scientific standards, of the individual contributions. If the resulting reports can be forwarded to us (series editors or Springer) this is very helpful. If no reports are forwarded or if other questions remain unclear in respect of homogeneity etc, the series editors may wish to consult external referees for an overall evaluation of the volume. A final decision to publish can be made only on the basis of the complete manuscript; however a preliminary decision can be based on a pre-final or incomplete manuscript. The strict minimum amount of material that will be considered should include a detailed outline describing the planned contents of each chapter.

    Volume editors and authors should be aware that incomplete or insufficiently close to final manuscripts almost always result in longer evaluation times. They should also be aware that parallel submission of their manuscript to another publisher while under consideration for LNM will in general lead to immediate rejection.

4. Manuscripts should in general be submitted in English. Final manuscripts should contain at least 100 pages of mathematical text and should always include
   - a general table of contents;
   - an informative introduction, with adequate motivation and perhaps some historical remarks: it should be accessible to a reader not intimately familiar with the topic treated;
   - a global subject index: as a rule this is genuinely helpful for the reader.

   Lecture Notes volumes are, as a rule, printed digitally from the authors' files. We strongly recommend that all contributions in a volume be written in the same LaTeX version, preferably LaTeX2e. To ensure best results, authors are asked to use the LaTeX2e style files available from Springer's web-server at

   ftp://ftp.springer.de/pub/tex/latex/svmonot1/ (for monographs) and

   ftp://ftp.springer.de/pub/tex/latex/svmultt1/ (for summer schools/tutorials).

   Additional technical instructions, if necessary, are available on request from:
   lnm@springer.com.

5. Careful preparation of the manuscripts will help keep production time short besides ensuring satisfactory appearance of the finished book in print and online. After acceptance of the manuscript authors will be asked to prepare the final LaTeX source files and also the corresponding dvi-, pdf- or zipped ps-file. The LaTeX source files are essential for producing the full-text online version of the book. For the existing online volumes of LNM see:

   http://www.springerlink.com/openurl.asp?genre=journal&issn=0075-8434.

   The actual production of a Lecture Notes volume takes approximately 12 weeks.

6. Volume editors receive a total of 50 free copies of their volume to be shared with the authors, but no royalties. They and the authors are entitled to a discount of 33.3 % on the price of Springer books purchased for their personal use, if ordering directly from Springer.

7. Commitment to publish is made by letter of intent rather than by signing a formal contract. Springer-Verlag secures the copyright for each volume. Authors are free to reuse material contained in their LNM volumes in later publications: a brief written (or e-mail) request for formal permission is sufficient.

Addresses:

Professor J.-M. Morel, CMLA,
École Normale Supérieure de Cachan,
61 Avenue du Président Wilson, 94235 Cachan Cedex, France
E-mail: morel@cmla.ens-cachan.fr

Professor B. Teissier, Institut Mathématique de Jussieu,
UMR 7586 du CNRS, Équipe "Géométrie et Dynamique",
175 rue du Chevaleret, 75013 Paris, France
E-mail: teissier@math.jussieu.fr

For the "Mathematical Biosciences Subseries" of LNM:

Professor P. K. Maini, Center for Mathematical Biology,
Mathematical Institute, 24-29 St Giles,
Oxford OX1 3LP, UK
E-mail : maini@maths.ox.ac.uk

Springer, Mathematics Editorial I,
Tiergartenstr. 17,
69121 Heidelberg, Germany,
Tel.: +49 (6221) 4876-8259
Fax: +49 (6221) 4876-8259
E-mail: lnm@springer.com